T0213016

Lecture Notes in Computer Science 10230

Commenced Publication in 1973
Founding and Former Series Editors:
Gerhard Goos, Juris Hartmanis, and Jan van Leeuwen

More information about this series at http://www.springer.com/series/7407

Timos Sellis · Konstantinos Oikonomou (Eds.)

Algorithmic Aspects of Cloud Computing

Second International Workshop, ALGOCLOUD 2016
Aarhus, Denmark, August 22, 2016
Revised Selected Papers

 Springer

Editors
Timos Sellis (iD)
Computer Science and Software Engineering
Swinburne University of Technology
Hawthorn, VIC
Australia

Konstantinos Oikonomou
Informatics
Ionian University
Kerkyra
Greece

ISSN 0302-9743   ISSN 1611-3349   (electronic)
Lecture Notes in Computer Science
ISBN 978-3-319-57044-0   ISBN 978-3-319-57045-7   (eBook)
DOI 10.1007/978-3-319-57045-7

Library of Congress Control Number: 2017937694

LNCS Sublibrary: SL1 – Theoretical Computer Science and General Issues

Printed on acid-free paper

This Springer imprint is published by Springer Nature
The registered company is Springer International Publishing AG
The registered company address is: Gewerbestrasse 11, 6330 Cham, Switzerland

Preface

The International Workshop on Algorithmic Aspects of Cloud Computing (ALGO-CLOUD) is an annual event aiming to tackle the diverse new topics in the emerging area of algorithmic aspects of computing and data management in the cloud. The increasing adoption of cloud computing introduces a variety of parallel and distributed algorithmic models and architectures. To leverage elastic cloud resources, scalability has to be a fundamental architectural design trait of new cloud databases. This challenge is manifested in new data models ("NoSQL"), replication, caching and partitioning schemes, relaxed consistency, and transaction guarantees as well as new protocols, APIs, indexing and storage services.

The aim of the workshop is to bring together researchers and practitioners in cloud computing algorithms, service design, and data architectures to exchange ideas and contribute to the development of this exciting and emerging new field.

ALGOCLOUD welcomes theoretical, experimental, methodological and application papers. Demonstration papers and high-quality survey papers are also welcome. As such, contributions are expected to span a wide range of algorithms for modeling, constructing, and evaluating operations and services in a variety of systems, including (but not limited to) virtualized infrastructures, cloud platforms, data centers, mobile ad hoc networks, peer-to-peer and grid systems, HPC architectures, etc.

Topics of interest addressed by this workshop include, but are not limited to:

- Algorithmic aspects of elasticity and scalability for distributed, large-scale data stores (e.g., NoSQL and columnar databases)
- Search and retrieval algorithms for cloud infrastructures
- Monitoring and analysis of elasticity for virtualized environments
- NoSQL, schemaless data modeling, integration
- Caching and load-balancing
- Storage structures and indexing for cloud databases
- New algorithmic aspects of parallel and distributed computing for cloud applications
- Scalable machine learning, analytics and data science
- High availability, reliability, failover
- Transactional models and algorithms for cloud databases
- Query languages and processing, programming models
- Consistency, replication and partitioning CAP, data structures and algorithms for eventually consistent stores

ALGOCLOUD 2016 took place on August 22, 2016, at the Lakeside Lecture Theatres (Søauditorierne) at Aarhus University, Denmark. It was collocated and was part of ALGO 2016 (August 22–26, 2016), the major annual congress that combines the premier algorithmic conference European Symposium on Algorithms (ESA) and a number of other specialized conferences and workshops, all related to algorithms and their applications, making ALGO the major European event for researchers, students, and practitioners in algorithms.

ALGOCLOUD 2016 was hosted by Aarhus University and sponsored by the Danish National Research Foundation Center MADALGO affiliated with the Department of Computer Science, and was organized in cooperation with the European Association for Theoretical Computer Science (EATCS).

The Program Committee (PC) of ALGOCLOUD 2016 was delighted by the positive response to the call for papers. The diverse nature of papers submitted demonstrated the vitality of the algorithmic aspects of cloud computing. All submissions underwent the standard peer-review process and were reviewed by at least three PC members, sometimes assisted by external referees. The PC decided to accept 11 original research papers that were presented at the workshop.

The program of ALGOCLOUD 2016 was complemented with a highly interesting tutorial, entitled "Big Data Management and Scalable Data Science: Key Challenges and (Some) Solutions," which was delivered by Prof. Volker Markl (Technische Universität Berlin, Germany). We wish to express our sincere gratitude to this distinguished scientist for the excellent tutorial he provided.

We hope that these proceedings will help researchers to understand and be aware of state-of-the-art algorithmic aspects of cloud computing, and that they will stimulate further research in the domain of algorithmic approaches in cloud computing in general.

September 2016   Timos Sellis
                                                Konstantinos Oikonomou

Organization

Steering Committee

Peter Triantafillou	University of Glasgow, UK
Spyros Sioutas	Ionian University, Greece
Christos Zaroliagis	University of Patras, Greece

Program Committee

Christos Anagnostopoulos	University of Glasgow, UK
Albert Bifet	University of Waikato, New Zealand
Alexis Delis	University of Athens, Greece
Marios Dikaiakos	University of Cyprus, Cyprus
Schahram Dustdar	Technical University of Vienna, Austria
Anastasios Gounaris	AUTH, Greece
Seif Haridi	Royal Institute of Technology, Sweden
Ioannis Karydis	Ionian University, Greece
Yannis Konstantinou	NTUA, Greece
Konstantinos Oikonomou	Ionian University, Greece
George Pallis	University of Cyprus, Cyprus
Apostolos Papadopoulos	AUTH, Greece
Timos Selis	Swinburne University of Technology, Australia
Spyros Sioutas	Ionian University, Greece
Peter Triantafillou	University of Glasgow, UK
Hong-Linh Truong	Technical University of Vienna, Austria
Dimitrios Tsoumakos	Ionian University, Greece
Michail Vassilakopoulos	UTH, Greece
Spyros Voulgaris	University of Patras, Greece

Additional Reviewers

Marios Kendea
Andreas Kosmatopoulos
Nikolaos Nodarakis
Athanasios Naskos

Big Data Management and Scalable Data Science: Key Challenges and (Some) Solutions (Tutorial)

Volker Markl

Technische Universität Berlin, Berlin, Germany

Abstract. The shortage of qualified data scientists is effectively limiting Big Data from fully realizing its potential to deliver insight and provide value for scientists, business analysts, and society as a whole. Data science draws on a broad number of advanced concepts from the mathematical, statistical, and computer sciences in addition to requiring knowledge in an application domain. Solely teaching these diverse skills will not enable us to on a broad scale exploit the power of predictive and prescriptive models for huge, heterogeneous, and high-velocity data. Instead, we will have to simplify the tasks a data scientist needs to perform, bringing technology to the rescue: for example, by developing novel ways for the specification, automatic parallelization, optimization, and efficient execution of deep data analysis workflows. This will require us to integrate concepts from data management systems, scalable processing, and machine learning, in order to build widely usable and scalable data analysis systems. In this talk, I will present some of our research results towards this goal, including the Apache Flink open-source big data analytics system, concepts for the scalable processing of iterative data analysis programs, and ideas on enabling optimistic fault tolerance.

Contents

Software Tools and Distributed Architectures for Cloud-Based Data Management

NSM-Tree: Efficient Indexing on Top of NoSQL Databases

Ioannis Kokotinis[1], Marios Kendea[1], Nikolaos Nodarakis[1(✉)], Angeliki Rapti[1],
Spyros Sioutas[2], Athanasios K. Tsakalidis[1], Dimitrios Tsolis[3],
and Yannis Panagis[4]

[1] Computer Engineering and Informatics Department,
University of Patras, 26504 Patras, Greece
{kokotinis,kendea,nodarakis,arapti,tsak}@ceid.upatras.gr
[2] Department of Informatics, Ionian University, 49100 Corfu, Greece
sioutas@ionio.gr
[3] Department of Cultural Heritage, Management and New Technologies,
University of Patras, 26504 Patras, Greece
dtsolis@upatras.gr
[4] Centre of Excellence for International Courts,
University of Copenhagen, 1455 Copenhagen, Denmark
ioannis.panagis@jur.ku.dk

Abstract. During the last years, there is a huge proliferation in the usage of location-based services (LBSs), mostly through a multitude of mobile devices (GPS, smartphones, mapping devices, etc.). The volume of the data derived by such services, grows exponentially and conventional databases tend to be ineffective in storing and indexing them efficiently. Ultimately, we need to turn to scalable solutions and methods using the NoSQL database model. Quite a few indexing methods exist in literature that work on top of NoSQL database. In this spirit, we deploy a new distributed indexing structure based on M-tree and perform a thorough experimental analysis to display its benefits.

Keywords: Big data · Distributed index · Range queries · M-tree · MapReduce · NoSQL

1 Introduction

With the recent advances in GPS-enabled devices, many applications extend their functionality in order to provide geolocation features to the user. Inevitably, geographic data (or spatiotemporal data) are all around us and are produced in an enormous rate with the extensive usage of these LBSs. Such data can be proved very useful when building geographic information systems (GIS) or applications (e.g. an application that retrieves the nearest restaurants based on some user-provided parameters). Until now geographic data are stored and indexed in conventional SQL databases, which also offer many tools and methods to support location queries. However, as the data volume continues expanding

© Springer International Publishing AG 2017
T. Sellis and K. Oikonomou (Eds.): ALGOCLOUD 2016, LNCS 10230, pp. 3–14, 2017.
DOI: 10.1007/978-3-319-57045-7_1

traditional databases and centralized indexing schemes lack in efficiency when processing queries. Inevitably, the flourish of NoSQL databases has led to a widespread development of many new indexing structures to efficiently process queries on large-scale data.

A database indexing scheme refers to a data structure that stores on a table a small amount of information, and makes easy to retrieve any piece of database information is requested in the form of a query. Usually, this table contains the ids of others' table records with appropriate references. A database driver may utilize the indexing structure for fast information access. The indexing scheme minimizes the query time as we do not need to make a full table scan to find the searched term. In order to create a database index we need to read the whole data collection from the disk and start building the structure on the main memory. If the structure is too big to fit in memory we transfer it to the disk and continue the construction as before. When this iterative process finishes, we merge the different parts of the index stored on the disk into a final bigger structure. The process to build an index over a distributed system resembles the previous one. Each node of the system creates a fraction of the structure and a unification process follows to form the final index. A suitable model to perform such a process is MapReduce [5]. In the Map task we read the input data and create the partial indexes while in the Reduce task we merge the partial indexes to the final unified index scheme [10].

In this paper, we focus on implementing the popular M-Tree index [4] on top of HBase [7,14] with the Hadoop framework [13,18], which is the open-source implementation of MapReduce. We call the new indexing scheme NSM-Tree (NoSQL M-Tree) and create it using a MapReduce process. We then proceed to an extensive experimental evaluation and compare NSM-Tree with the original M-Tree to display the gains of the new structure. The rest of the paper is organized as follows: in Sect. 2 we present some preliminaries and discuss the related work. In Sect. 3 we describe the implementation of NSM-Tree while in Sect. 4 we evaluate the structure considering a number of different parameters. Finally, in Sect. 5 we conclude the paper and discuss future directions.

2 Preliminaries

2.1 Previous Work

The research around the database indexing domain is quite extensive. Early studies focus on creating robust and efficient structures for answering range queries. One popular structure is the segment tree [6] which stores intervals (segments) and allows querying which of the stored segments contain a given point. It is very efficient and for a set of n intervals uses $O(n \log n)$ space and $O(n \log n)$ time to build it. It supports searching for intervals in $O(\log n + k)$ time where k is the number of intervals retrieved. Another efficient structure is the R-Tree [8] whose primary use is the indexing of multi-dimensional information such as geographical coordinates, rectangles or polygons. Two modifications of

R-Tree which offer better query time are the R+-Tree [16] and R*-Tree [1] structures. M-Tree data structure [4] displays a similar logic with the R-Tree and its derivatives with the difference that it indexes circular areas instead of rectangles. Thus, it is ideal for k-nearest neighbor (kNN) queries. A quad tree [12] is a tree data structure in which each internal node has exactly four children. Quad trees are most often used to partition a two-dimensional space by recursively subdividing it into four quadrants or regions. An extension of quad tree in d dimensions constitutes the k-d tree [2].

The methods proposed above can handle data of small size in one or more dimensions, thus their use is limited in centralized environments only. During the recent years, the researchers have focused on developing approaches that are applicable in distributed environments, like our method, and can manipulate big data in an efficient manner. The MapReduce framework seems to be suitable for processing such queries. For instance, the authors in [17] present the MRST-Tree which is the MapReduce distributed version of segment tree which greatly speeds up the native HBase and Hive [3,15] support for interval queries. An older distributed version of the same structure is the DST (Distributed Segment Tree) [19], which in essence is a layered distributed hash table (DHT) structure that incorporates the concept of segment tree. In [11] the authors propose a parallel and distributed implementation of quad and k-d tree when managing data produced by LBSs. Moreover, in [9] authors describe the new structure HB+-Tree, which is an extension of the original B+-Tree adapted to a distributed environment. In the context of this paper, we propose the NSM-Tree which is the MapReduce version of M-Tree built on top of HBase. We compare the new structured with its centralized predecessor to get an insight of the benefits of NSM-Tree.

2.2   M-Tree Overview

The M-Tree is a dynamic tree structure (generalization of B-tree) and is mainly used for indexing geospatial data. The function that calculates the distance between two points in space adheres to the theorems of symmetry and triangular inequality. The search algorithm of M-Tree is implemented as a binary search algorithm in each dimension, and is suitable for range queries in the form (x, y), where $min_x \leq x \leq max_x$ and $min_y \leq y \leq max_y$.

The data stored in M-Tree is a set of the form $O = (key, value)$, where key is the latitude and longitude of a point and $value$ contains information of this location. The dataset is organized in metric spaces $MS = (D, d)$, where D is the center of a circle with radius d. Each node can hold up to M records and can be distinguished in two types: the leafs node that contain the sorted data, and the inner nodes that contain indexes to the actual data according to their value. For each sorted object in a leaf node there exist a record $entry(O_j) = [O_j, oid(O_j), d(O_j, P(O_j))]$, where $oid(O_j)$ is the unique id of the record and $d(O_j, P(O_j))$ is the Euclidean distance of object O_j from his parent $P(O_j)$. The records of the inner nodes are of the form $entry(O_r) = [O_r, ptr(T(O_r)), r(O_r), d(O_r, P(O_r))]$, where $r(O_r) > 0$ and is the radius of the area covered by this object, and $ptr(T(O_r))$ points to the root of subtree $T(O_r)$ (the tree that covers the object O_r). The radius $r(O_r)$ always satisfies the inequality $d(O_j, O_r) \leq r(O_r)$, for each O_j that belongs to the tree $T(O_r)$ defined by the object O_r [4].

2.3   MapReduce Model and NoSQL Databases

Here, we briefly describe the MapReduce model [5] and NoSQL databases [7]. The data processing in MapReduce is based on input data partitioning; the partitioned data is processed by a number of tasks executed in many distributed nodes. There exist two major task categories called *Map* and *Reduce* respectively. Given input data, a *Map* function processes the data and outputs key-value pairs. Based on the Shuffle process, key-value pairs are grouped and then each group is sent to the corresponding Reduce task. A user can define his own Map and Reduce functions depending on the purpose of his application. The input and output formats of these functions are simplified as key-value pairs. Using this generic interface, the user can focus on his own problem and does not have to care how the program is executed over the distributed nodes. The architecture of MapReduce model is depicted in Fig. 1.

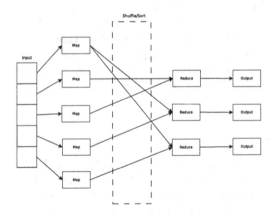

Fig. 1. Architecture of MapReduce model

Relational database management systems (RDBMSes) have typically played an integral role when designing and implementing business applications. As soon as you have to retain information about your users, products, sessions, orders, and so on, you are typically going to use some storage backend providing a persistence layer for the frontend application server. This works well for a limited number of records, but with the dramatic increase of data being retained, some of the architectural implementation details of common database systems show signs of weakness. Over the past four or five years, the pace of innovation to fill that exact problem space has gone from slow to insanely fast. It seems that every week another framework or project is announced to fit a related need. This caused the advent of the so-called NoSQL solutions. The term quickly rose to fame as there was simply no other name for this new class of products. The difference between RDBMSes and NoSQL is on a lower level, especially when it comes to schemas or ACIDlike transactional features, but also regarding the actual storage architecture. A lot of these new kinds of systems do one thing first: throw out the limiting factors in truly scalable systems (e.g. database demormalization).

3   NSM-Tree Implementation

In this subsection we describe the NoSQL version of M-Tree (NSM-Tree) as it was implemented in the context of this work. The structure is built on top of HBase, which is a column-oriented database model for storing big data. This means that the data are stored in columns and the rows are distinguished by a unique key. We describe the new index and use pseudo-code wherever needed.

In the original M-Tree the construction follows a top-down design. The objects are inserted to a node N until it overflows, which means that the number of inserted objects is greater than a threshold M. In this case we use a split routine. The split function sends two route objects of node N to its parent, partitions the records into two different sets and stores the one set to N and the other set to a new node N'. As a result the objects in the original M-Tree may appear to more than one nodes.

In our NSM-Tree implementation over HBase, we keep for each row in the table, all the appearances of an object on different nodes. The appearance order is from the root to the leafs. The key for each row is the longitude and latitude delimited by a semicolon and the columns contain the whole information of the tree. We maintain two column families, cf_1 and cf_2. Family cf_1 consists of four columns: (1) column *areaInfo* which declares the value of a point, (2) column *distanceFromParent* which records the Euclidean distance of the object from its parent, (3) column *parent* which stores the row key of the parent of this object, and (4) column *radius* which declares the maximum Euclidean distance between the object and its children. Family cf_2 contains as many columns as the size M of M-Tree, and stores the keys of the objects that are located in the root of the subtree each object points to.

As mentioned before, an object may appear at multiple nodes of the tree, thus the table columns should store this information. We achieve this by keeping in the value field, the node ids delimited by an exclamation mark. Consider e.g. Figure 2 and the row with key '0; 1'. Column *parent* contains the keys of all the parents of the objects with key '0; 1'. If we split this value on exclamation mark we get the values *root*, '0; 1' and '0; 1', which means that the root node contains the key '0; 1' and the same value is propagated into two nodes down the tree. Since the parents of the object with key '0; 1' are objects with the same key or the key is located at the root node, the values of column *distanceFromParent* are zero. Examining column *radius*, we observe that the first object with key '0; 1' in the tree has radius equal to 9.48, the second has radius equal to 4.47 and the last one is located at a leaf node.

ROW	COLUMN+CELL
0;1	column=cf1:areaInfo, timestamp=1439111353847, value=A H Watwood Elementary
School;school;AL;Talladega;01;121;;;Childersburg;TIFF;TFW;\xC2\xA0;TIGER;	
0;1	column=cf1:distanceFromParent, timestamp=1439111353847, value=0.0!0.0!0.0!
0;1	column=cf1:parent, timestamp=1439111353847, value=root!0;1!0;1!
0;1	column=cf1:radius, timestamp=1439111353847, value=9.48683298050513814.4721359549995814leaf!
0;1	column=cf2:ptr1, timestamp=1439111353847, value=3;-8!0;1!leaf!
0;1	column=cf2:ptr2, timestamp=1439111353847, value=0;1!4;-1!leaf!
1;-8	column=cf1:areaInfo, timestamp=1439111353847, value=A C Moore Elementary
School;school;AL;Escambia;01;053;;;Atmore;TIFF;TFW;\xC2\xA0;TIGER;	
1;-8	column=cf1:distanceFromParent, timestamp=1439111353847, value=2.0!
1;-8	column=cf1:parent, timestamp=1439111353847, value=3;-8!
1;-8	column=cf1:radius, timestamp=1439111353847, value=leaf!
1;-8	column=cf2:ptr1, timestamp=1439111353847, value=leaf!
1;-8	column=cf2:ptr2, timestamp=1439111353847, value=leaf!

Fig. 2. Sample from HBase table of NSM-Tree

As far as cf_2 is concerned, first we store the value of M for the NSM-Tree that results from the number of columns the family has (in Fig. 2, $M = 2$). The information of this column family is extracted by reading all the columns at the same time. Thus, for the objects with key 0; 1 we get that the first one points at a subtree the root of which contains the objects '3; −8' and '0; 1', the second points at a subtree root with the objects '0; 1' and '4; −1' and the third, as mentioned before, is a leaf. From the above information we can also verify the radius values (for example $d((0;1),(3;-8) = 9.48 > d((0;1),(0;1)) = 0$).

Next, we describe the construction of NSM-Tree in HBase and the way the queries are performed using pseudo-code. The data are initially stored in HDFS and use a centralized method to create a serialized file in HDFS. Then, we use a Map-only job to read the serialized file and create the NSM-Tree in HBase. Another way of doing this and save time in the tree construction, is to build the tree directly in HBase as a set of smaller subtrees. However, this process was very complex and requires much more nodes than we had available (see below in the Experimental Evaluation section) to fully understand its benefits. Hence, we leave it for future work. The range search in HBase is similar to the original search routine of M-Tree. The algorithm works by tokenizing the information stored in HBase and getting a list of the values between the ?!? characters for each column. We also count the number of times each HBase row was visited by the search algorithm, so that we can get the required information from the above lists in a quick and easy manner, without having to read the entire information that is stored in each column, each time the search algorithm demands it. We outline below the pseudo-codes for each operation apart from range search algorithm, which is described in [4]. It is important to note that in *BuildNSMT* algorithm the *Setup* function is executed only once in each node before the *Map* function executions.

It is obvious that the range search algorithm will be quite faster in NSM-Tree than in M-Tree, something that is confirmed to the experimental evaluation. The reason is that in NSM-Tree the search algorithm is performed in a distributed

---

1: **function** CREATESERIALIZEDFILE($inputPath, outputPath, M$)
2:     Create root node R;
3:     $T \leftarrow$ createMTree($R, inputPath$);
4:     $C \leftarrow T$.getRootChilden();
5:     T.saveToSerFile($outputPath$);
6:     Create HBase table HT with cf_1 and cf_2;
7:     HT.put(C);
8:     BuildNSMT(M);
9: **end function**

BuildNSMT(M)

1: **function** SETUP($inputPath, M$)
2:     $treeSize \leftarrow M$;
3:     $tree \leftarrow$ readFile($inputPath$); // Build tree in memory
4: **end function**

5: **function** MAP($k1, v1$)
6:     $key \leftarrow$ getKey($v1$); $val \leftarrow$ getVal($v1$);
7:     Get children of node with key and store them in list L;
8:     Create array A of size $treeSize$ to store children information;
9:     **for all** $v \in L$ **do**
10:         Get distance d, parent p, radius r;
11:         Fill A with children keys containing this object;
12:         $B \leftarrow < d, p, r, A >$;
13:         HT.put(B);
14:     **end for**
15: **end function**

1: **function** RUNSEARCH($HT, key, range$)
2:     $root \leftarrow$ getTableRoot(HT);
3:     $visits \leftarrow$ calculateVisits(key);
4:     $info \leftarrow$ tokenizeRow(key);
5:     Execute range search algorithm for key, range and $info(visits)$;
6: **end function**

---

manner among the nodes of the cluster. Each node searches into a subtree of NSM-tree and as a result the query time gains a significant speedup.

4 Experimental Evaluation

In this section, we conduct a series of experiments to evaluate the performance of our method under many different perspectives. More precisely, we take into consideration the value of M and different ranges for the queries. Our cluster consists of 2 computing nodes (2 VMS, 1 master and 1 slave) each one of which has two 2.1 GHz CPU processors, 4 GB of memory, 40 GB hard disk and the nodes are connected by 1 gigabit Ethernet. On each node, we install the Debian

7.4 operating system, Java 1.7.0 with a 64-bit Server VM, Hadoop 1.0.4 and HBase 0.94.5. In order to adapt the Hadoop environment to our application, we apply the following changes to the default Hadoop configurations: the replication factor is set to 1, the maximum number of Map and Reduce tasks in each node is set to 3, the DFS chunk size is set to 64 MB and the size of virtual memory for each Map and Reduce task to 512 MB.

For the purposes of evaluating the original M-Tree against the proposed NSM-Tree, we used a real dataset from Libre Map Project[1], which consists of locations of various points of interest distributed in USA. The data are initially stored in HTML tables and we use a crawler written in Python to extract them into a semicolon-delimited file. We ended up in a file with 1.677.030 records which is then transferred to the cluster HDFS. We performed a comparative analysis between the two structures in terms of query response time, and we did not take into account the time needed to construct the tree, since it is an one time process. Furthermore, we assigned values to the parameter M from the set $10, 20, 30, 40, 50$.

We conducted three different kind of experiments. In the first experiment, we measured the query response time when we search a random element for a random search range. We assured that the range is large enough to return a good proportion of the dataset without having to scan the whole tree. In the second experiment, we searched for 1000 random keys with constant range. Finally, we searched for a random constant element with 1000 random search ranges. In all experiments, we recorded the fluctuation of response time as M changes. In Fig. 3, we present the results for the first experiment using normal and logarithmic scale for the response time. Observe that for smaller values of M (until $M = 30$), the two approaches present almost similar behavior. For $M > 30$, NSM-Tree is by an order of magnitude faster in retrieving the results. Thus, we conclude that for a random search the distributed solution is at least as much efficient as M-Tree for small M, while it significantly outperforms the centralized solution as M gets bigger.

In Fig. 4, the outcome of the second experiment is displayed. According to the charts, M-Tree introduces a steady behavior which translates to almost the same response time to answer 1000 queries, regardless of the value of M. On the contrary, the curve for NSM-Tree response time, declines as M rises. More concretely, especially by examining the logarithmic scale graph, the distributed solution exhibits a great speedup in the query process as M increases, compared to the centralized solution whose curve resembles a straight line.

Figure 5 presents the execution time graphs for various values of M and for the third and final experiment. The outcome is similar to the one in Fig. 4. The only difference here is that the query response time is larger because the requested object is always in the search range of the query. As a consequence, the query always returns results, which is not always the case in the second experiment. It is also worth noting that in Figs. 4 and 5 the NSM-Tree needs more time to answer the queries for $M = 30$ than $M = 20$, a result that contradicts the theory.

---

[1] Map and GIS Data By US State: http://libremap.org/data/.

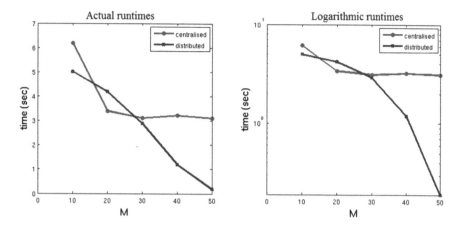

Fig. 3. Response time of M-Tree and NSM-Tree for 1 random search and different values of M

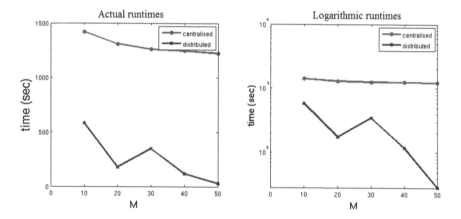

Fig. 4. Response time of M-Tree and NSM-Tree for 1000 random searches with constant range and different values of M

A possible explanation is that this result occurs because of the way the points are distributed in the index.

The results from Figs. 3, 4 and 5 are summarized in Tables 1 and 2 in order to obtain a better overall picture about the behavior of centralized and distributed solutions. We observe that for the centralized solution the query processes need less time to complete, as M increases, but the differences are relatively small.

The results of NSM-Tree in Table 2 display the same tendency with the respective outcome of M-Tree in Table 1. The main difference is that the decrease rate of query time, as M rises, follows an exponential trend. These two tables, along with Figs. 3, 4 and 5, prove the supremacy of NSM-Tree over M-Tree even for a cluster with a small number of nodes. The divergence in response time

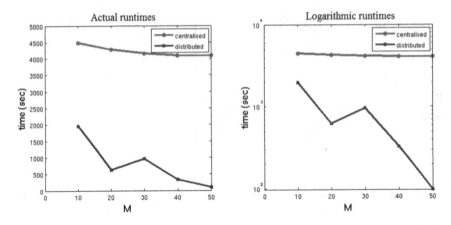

Fig. 5. Response time of M-Tree and NSM-Tree for 1000 range searches of one element and different values of M

Table 1. M-Tree response times for various combinations of query and M

M	1 element - 1 range	1000 elements - 1 range	1 element - 1000 ranges
10	6.19 s	1420.14 s	4486.04 s
20	3.39 s	1309.57 s	4280.52 s
30	3.10 s	1258.62 s	4159.44 s
40	3.21 s	1245.58 s	4101.17 s
50	3.11 s	1225.42 s	4008.93 s

Table 2. NSM-Tree response times for various combinations of query and M

M	1 element - 1 range	1000 elements - 1 range	1 element - 1000 ranges
10	5.02 s	580.45 s	1958.18 s
20	4.19 s	175.79 s	620.82 s
30	2.89 s	347.77 s	969.91 s
40	1.19 s	118.19 s	328.25 s
50	0.19 s	28.53 s	98.19 s

is expected to increase for larger data volumes and infrastructures with larger node cardinality.

To sum up the experimental evaluation, as an overall observation for both structures we can note that the query response time is inversely proportional to the increase of the node size M. This happens because larger values of M lead to smaller tree depth. Consequently, the search algorithm will traverse a smaller path from the root to the leafs to obtain the query answer. Moreover, the NSM-Tree overperforms M-Tree, especially as M grows. It is worth noting that the processing a query consisting of an object with 1000 random ranges is more time consuming compared to the search of 1000 objects with a constant range. The explanation lies in the configuration parameters we chose. When searching for 1000 objects, we defined the location of these objects to be close to the geographical coverage area of the dataset. If an object lies outside the coverage area, the search process stops almost immediately (looks only in the root node) if the search range is not big enough to include at least one tree element. On the other hand, when searching for an object with 1000 random ranges we defined the element to be inside the geographical coverage area of the dataset, in order to never have an empty result set.

5   Conclusions and Future Work

In the context of this work, we presented NSM-Tree which is the distributed NoSQL version of M-Tree. We described the way the structure is built on top of Hadoop and HBase using pseudo-code and evaluated its performance with the original centralized M-Tree index. Our solutions proves its superiority compared to its centralized predecessor even for a small cluster infrastructure.

Although, to our best knowledge, this is the first attempt to build a cloud-based M-Tree a lot can be done to improve its performance. Some future steps may include an evaluation over a significantly larger cluster with higher volume data, the fully distributed construction of the tree as well as the implementation of an update algorithm when new elements are inserted in the database. Finally, it is necessary to test the index under heavy query load and explore fault tolerance issues.

Acknowledgements. This research was kindly supported by the C. Carathéodory Research Program at University of Patras, Greece.

References

1. Beckmann, N., Kriegel, H.-P., Schneider, R., Seeger, B.: The R*-Tree: an efficient and robust access method for points and rectangles. In: Proceedings of the 1990 ACM SIGMOD International Conference on Management of Data, pp. 322–331. ACM, New York (1990)
2. Bentley, J.L.: Multidimensional binary search trees used for associative searching. Commun. ACM **18**, 509–517 (1975)
3. Capriolo, E., Wampler, D., Rutherglen, J.: Programming Hive Data Warehouse and Query Language for Hadoop. O'Reilly Media, Sebastopol (2012)
4. Ciaccia, P., Patella, M., Zezula, P.: M-Tree: an efficient access method for similarity search in metric spaces. In: Proceedings of the 23rd International Conference on Very Large Data Bases, pp. 426–435. Morgan Kaufmann Publishers Inc., San Francisco (1997)
5. Dean, J., Ghemawat, S.: MapReduce: simplified data processing on large clusters. In: Proceedings of the 6th Symposium on Operating Systems Design and Implementation, pp. 137–150. USENIX Association, Berkeley (2004)
6. Berg, M.D., Cheong, O., Van Kreveld, M., Overmars, M.: Computational Geometry: Algorithms and Applications, 3rd edn. Springer, Heidelberg (2008)

7. George, L.: HBase: The Definitive Guide. O'Reilly Media, Sebastopol (2011)
8. Guttman, A.: R-Trees: a dynamic index structure for spatial searching. In: Proceedings of the 1984 ACM SIGMOD International Conference on Management of Data, pp. 47–57. ACM, New York (1984)
9. Kaplanis, A., Kendea, M., Sioutas, S., Makris, C., Tzimas, G.: HB+Tree: Use Hadoop and HBase even your data isn't that big. In: Proceedings of the 30th Annual ACM Symposium on Applied Computing, pp. 973–980. ACM, New York (2015)
10. McCreadie, R., McDonald, C., Ounis, I.: Comparing distributed indexing: to mapreduce or not? In: Proceedings of the 7th Workshop on Large-Scale Distributed Systems for Information Retrieval, pp. 41–48. CEUR Workshop Proceedings (2009)
11. Nishimura, S., Das, S., Agrawal, D., Abbadi, A.E.: MD-HBase: a scalable multidimensional data infrastructure for location aware services. In: Proceedings of the IEEE 12th International Conference on Mobile Data Management, vol. 1, pp. 7–16. IEEE Computer Society, Washington, DC (2011)
12. Samet, H., Webber, E.R.: Storing a collection of polygons using quadtrees. ACM Trans. Graph. **4**, 182–222 (1985)
13. The Apache Software Foundation: Hadoop homepage. http://hadoop.apache.org/
14. The Apache Software Foundation: HBase homepage. http://hbase.apache.org/
15. The Apache Software Foundation: Hive homepage. http://hive.apache.org/
16. Sellis, K.T., Roussopoulos, N., Faloutsos, C.: The R+-Tree: a dynamic index for multi-dimensional objects. In: Proceedings of the 13th International Conference on Very Large Data Bases, pp. 507–518. Morgan Kaufmann Publishers Inc., San Francisco (1987)
17. Sfakianakis, G., Patlakas, I., Ntarmos, N., Triantafillou, P.: Interval indexing and querying on key-value cloud stores. In: Proceedings of the 29th IEEE International Conference on Data Engineering, pp. 805–816 (2013)
18. White, T.: Hadoop: The Definitive Guide, 3rd edn. O'Reilly Media/Yahoo Press (2012)
19. Zheng, C., Shen, G., Li, S., Shenker, S.: Distributed segment tree: support of range query and cover query over DHT. In: Proceedings of the 5th International Workshop on Peer-To-Peer Systems (2006)

An Apache Spark Implementation for Sentiment Analysis on Twitter Data

Alexandros Baltas, Andreas Kanavos(✉), and Athanasios K. Tsakalidis

Computer Engineering and Informatics Department,
University of Patras, Patras, Greece
{ampaltas,kanavos,tsak}@ceid.upatras.gr

Abstract. Sentiment Analysis on Twitter Data is a challenging problem due to the nature, diversity and volume of the data. In this work, we implement a system on Apache Spark, an open-source framework for programming with Big Data. The sentiment analysis tool is based on Machine Learning methodologies alongside with Natural Language Processing techniques and utilizes Apache Spark's Machine learning library, MLlib. In order to address the nature of Big Data, we introduce some pre-processing steps for achieving better results in Sentiment Analysis. The classification algorithms are used for both binary and ternary classification, and we examine the effect of the dataset size as well as the features of the input on the quality of results. Finally, the proposed system was trained and validated with real data crawled by Twitter and in following results are compared with the ones from real users.

Keywords: Apache Spark · Big Data · Classification · Microblogging · Sentiment Analysis · Social media analytics

1 Introduction

Nowadays people share moments, experiences and feelings through social networks. Microblogging platforms, namely Twitter, have recently become very popular. Founded in 2006, Twitter is a service which allows users to share 140-character posts. Having gained massive popularity while being widely considered one of the most influential services on the World Wide Web. Twitter has resulted in hosting massive datasets of information. Thus its data is gaining increasing interest. People use Twitter to share experiences and emotions with their friends about movies, products, events etc., so a system that extracts sentiments through an online community may have many real-life applications such as recommendation systems. This enormously continuous stream of Twitter data posts reflects the users opinions and reactions to phenomena from political events all over the world to consumer products [20]. It is well pointed that Twitter posts relate to the user's behavior and often convey substantial information about their emotional state [3].

© Springer International Publishing AG 2017
T. Sellis and K. Oikonomou (Eds.): ALGOCLOUD 2016, LNCS 10230, pp. 15–25, 2017.
DOI: 10.1007/978-3-319-57045-7_2

Unlike other networks, users' posts in Twitter have some special characteristics. The short length that the posts are allowed to have, results in more expressive emotional statements. Analyzing tweets and recognizing their emotional content is a very interesting and challenging topic in the microblogging area. Recently many studies have analyzed sentiment from documents or web-related content, but when such applications are focused on microblogging, many challenges occur. The limited size of the messages, along with the wide range of subjects discussed, make sentiment extraction a difficult process. Concretely, researchers have used long-known machine learning algorithms in order to analyze sentiments. So the problem of sentiment extraction is transformed into a classification problem. Datasets of classified tweets are used to train classifiers which in following are used to extract the sentiments of the messages.

In the meantime, as data grows, cloud computing evolves. Frameworks like Hadoop, Apache Spark, Apache Storm and distributed data storages like HDFS and HBase are becoming popular, as they are engineered in a way that makes the process of very large amounts of data almost effortless. Such systems evolve in many aspects, and as a result, libraries, like Spark's MLlib that make the use of Machine Learning techniques possible in the cloud, are introduced.

In this paper we aim on creating a Sentiment Analysis tool of Twitter data based on Apache Spark cloud framework, which classifies tweets using supervised learning techniques. We experiment with binary and ternary classification, and we focus on the change in accuracy caused by the training dataset size, as well as the features extracted from the input.

The remainder of the paper is structured as follows: Section 2 presents the related work. Section 3 presents cloud computing methodologies, while Sect. 4 presents the classification algorithms used in our proposed system. Section 5 presents the steps of training as well as the two types of classification, binary and ternary. Moreover, Sect. 6 presents the evaluation experiments conducted and the results gathered. Ultimately, Sect. 7 presents conclusions and draws directions for future work.

2   Related Work

In the last decade, there has been an increasing interest in studies of Sentiment Analysis as well as emotional models. This is mainly due to the recent growth of data available in the World Wide Web, especially of those that reflect people's opinions, experiences and feelings [17]. Sentiment Analysis is studied in many different levels. In [22], authors implement an unsupervised learning algorithm that classifies reviews, thus performing document level classification. In [13] authors operate in a word and sentence level, as they classify people's opinions. Moreover, Wilson et al. [24] operate on a phrase level, by determining the neutrality or polarity of phrases. Machine learning techniques are frequently used for this purpose. Pang et al. [18] used Naive Bayes, Maximum Entropy and SVM classifiers so as to analyze sentiment of movie reviews. Boiy and Moens [2] utilized classification models with the aim of mining the sentiment out of multilingual web texts.

Twitter data are used by researchers in many different areas of interest. In [8], Tweets referring to Hollywood movies are analyzed. They focused on classifying the Tweets and in following on analyzing the sentiment about the Hollywood movies in different parts of the world. Wang et al. [23] used a training data of 17000 Tweets in order to create a real-time Twitter Sentiment Analysis System of the U.S. 2012 Presidential Election Cycle. In addition, in [15], authors present a novel method for Sentiment Learning in the Spark framework; the proposed algorithm exploits the hashtags and emoticons inside a tweet, as sentiment labels, and proceeds to a classification procedure of diverse sentiment types in a parallel and distributed manner.

Other studies that investigate the role of emoticons on sentiment analysis of Tweets are the ones in [19,25]. In both works, Lexicons of Emoticons are used to enhance the quality of the results. Authors in [4] propose a system that uses an SVM classifier alongside a rule-based classifier so as to improve the accuracy of the system. In addition, in [3], authors utilized the Profile of Mood States psychometric method for analyzing Twitter posts and reached the conclusion that "the events in the social, political, cultural and economic sphere do have significant, immediate and highly specific effect on the various dimensions of public mood". Commercial companies and associations could exploit Twitter for marketing purposes, as it provides an effective medium for propagating recommendations through users with similar interests.

There is a lot of research interest in studying different types of information dissemination processes on large graphs and social networks. Naveed et al. [14] analyze tweet posts and forecast for a given post the likelihood of being retweeted on its content. Authors indicate that tweets containing negative emoticons are more likely to be retweeted than tweets with positive emoticons. Finally, previous works regarding emotional content are the ones in [9–12]; they presented various approaches for the automatic analysis of tweets and the recognition of the emotional content of each tweet based on Ekman emotion model, where the existence of one or more out of the six basic human emotions (Anger, Disgust, Fear, Joy, Sadness and Surprise) is specified.

3   Cloud Computing

3.1   MapReduce Model

MapReduce is a programming model which enables the process of large datasets on a cluster using a distributed and parallel algorithm [6]. A MapReduce program consists of 2 main procedures, Map() and Reduce() respectively, and is executed in 3 steps; Map, Shuffle and Reduce. In the Map phase, input data is partitioned and each partition is given as an input to a worker that executes the map function. Each worker processes the data and outputs key-value pairs. In the Shuffle phase, key-value pairs are grouped by key and each group is sent to the corresponding Reducer. Apache Hadoop is a popular open source implementation of the Map Reduce model.

3.2 Spark Framework

Spark Framework[1] is a newer framework built in the same principles as Hadoop. While Hadoop is ideal for large batch processes, it drops in performance in certain scenarios, as in iterative or graph based algorithms. Another problem of Hadoop is that it does not cache intermediate data for faster performance but instead, it flushes the data to the disk between each step. In contrast, Spark maintains the data in the workers' memory and as a result it outperforms Hadoop in algorithms that require many operations. Spark offers API in Scala, Java, Python and R and can operate on Hadoop or standalone while using HDFS, Cassandra or HBase.

3.3 MLlib

Spark's ability to perform well on iterative algorithms makes it ideal for implementing Machine Learning Techniques as, at their vast majority, Machine Learning algorithms are based on iterative jobs. MLlib[2] is Apache Spark's scalable machine learning library and is developed as part of the Apache Spark Project. MLlib contains implementations of many algorithms and utilities for common Machine Learning techniques such as Clustering, Classification, Regression.

4 Machine Learning Techniques

In this work, we utilized three classification algorithms in order to implement the Sentiment Analysis Tool. We examined both Binary and Ternary Classification on different datasets. On the Binary Classification case, we focus on the way that the dataset size affects the results, while on the Ternary Classification case, the focus is given on the impact of the different features of the feature vector given as an input to the classifier. The three algorithms utilized are Naive Bayes, Logistic Regression and Decision Trees.

4.1 Naive Bayes

Naive Bayes is a simple multiclass classification algorithm based on the application of Bayes' theorem. Each instance of the problem is represented as a feature vector, and it is assumed that the value of each feature is independent of the value of any other feature. One of the advantages of this algorithm is that it can be trained very efficiently as it needs only a single pass to the training data. Initially, the conditional probability distribution of each feature given class is computed, and then Bayes' theorem is applied to predict the class label of an instance.

---

[1] http://spark.apache.org/.
[2] http://spark.apache.org/mllib/.

4.2 Logistic Regression

Logistic regression is a regression model where the dependent variable can take one out of a fixed number of values. It utilizes a logistic function to measure the relationship between the instance class, and the features extracted from the input. Although widely used for binary classification, it can be extended to solve multiclass classification problems.

4.3 Decision Trees

The decision tree is a classification algorithm that is based on a tree structure whose leaves represent class labels while branches represent combinations of features that result in the aforementioned classes. Essentially, it executes a recursive binary partitioning of the feature space. Each step is selected greedily, aiming for the optimal choice for the given step by maximizing the information gain.

5 Implementation

The overall architecture of the proposed system is depicted in Fig. 1 taking into account the corresponding modules of our approach. Inititally, a pre-processing step, as shown in following subsection, is utilized and in following the classifiers for estimating the sentiment of each tweet, are used.

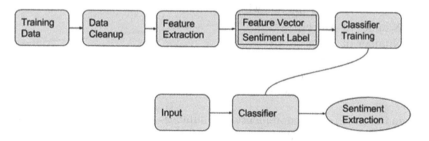

Fig. 1. Proposed system architecture

5.1 Binary Classification

For the Binary Classification, we used a dataset[3] of 1.578.627 pre-classified tweets as Positive or Negative. We split the original dataset into segments of 1.000, 2.000, 5.000, 10.000, 15.000, 20.000 and 25.000 tweets. Then for each segment, all metadata were discarded and each tweet was transformed to a vector of unigrams; unigrams are the frequencies of each word in the tweets.

---

[3] http://thinknook.com/twitter-sentiment-analysis-training-corpus-dataset-2012-09-22/.

5.2   Ternary Classification

Regarding Ternary Classification, we used two datasets[4] that were merged into one which eventually consists of 12.500 tweets. In the original datasets, each row contains the tweet itself, the sentiment, and other metadata related to the corresponding tweet. During the preprocessing, all irrelevant data were discarded, and we only used the actual text of the tweet, as well as the label that represents the sentiment; positive, negative or neutral.

Each tweet is then tokenized and processed. Occurrences of usernames and URLs are replaced by special tags and each tweet is finally represented as a vector which consists of the following features:

- **Unigrams**, which are frequencies of words occurring in the tweets.
- **Bigrams**, which are frequencies of sequences of 2 words occurring in the tweets.
- **Trigrams**, which are frequencies of sequences of 3 words occurring in the tweets.
- **Username**, which is a binary flag that represents the existence of a user mention in the tweet.
- **Hashtag**, which is a binary flag that represents the existence of a hashtag in the tweet.
- **URL**, which is a binary flag that represents the existence of a URL in the tweet.
- **POS Tags**, where we used the Stanford NLT MaxEnt Tagger [21] to tag the tokenized tweets and the following are counted:
  1. Number of Adjectives
  2. Number of Verbs
  3. Number of Nouns
  4. Number of Adverbs
  5. Number of Interjections

Then the ratios of the aforementioned numbers to the total number of tokens of each tweet are computed.

6   Evaluation

The results of our work are presented in the following Tables 1, 2, 3, 4 and 5. F-Measure is used as the evaluation metric of the different algorithms. For the binary classification problem (Table 1), we observe that Naive Bayes performs better than Logistic Regression and Decision Trees. It is also obvious that the dataset size plays a rather significant role for Naive Bayes, as the F-Measure value rises from 0.572 for a dataset of 1.000 tweets to 0.725 for the dataset of 25.000 tweets. On the contrary, the performance of Logistic Regression and Desicion Trees is not heavily affected by the amount of the tweets in the dataset.

---

[4] https://www.crowdflower.com/data-for-everyone/.

Table 1. Binary Classification - F-Measure

Dataset size	Naive Bayes	Logistic Regression	Decision Trees
1000	0.572	0.662	0.597
5000	0.684	0.665	0.556
10000	0.7	0.649	0.568
15000	0.71	0.665	0.575
20000	0.728	0.651	0.59
25000	0.725	0.655	0.56

Table 2. Ternary Classification - F-Measure

Classifier	Positive	Negative	Neutral	Total
Naive Bayes	0.717	0.75	0.617	0.696
Logistic Regression	0.628	0.592	0.542	0.591
Decision Trees	0.646	0.727	0.557	0.643

Table 3. Ternary Classification - F-Measure for Naive Bayes

Features	Positive	Negative	Neutral	Total
Complete feature vector	0.717	0.75	0.617	0.696
w/o Unigrams	0.628	0.602	0.537	0.592
w/o Bigrams	0.714	0.769	0.629	0.705
w/o Trigrams	0.732	0.77	0.643	0.716
w/o User	0.718	0.751	0.618	0.698
w/o Hashtag	0.721	0.739	0.608	0.692
w/o URL	0.72	0.748	0.619	0.697
w/o POS Tags	0.716	0.748	0.617	0.695

Regarding ternary classification, Naive Bayes outperforms the other two algorithms as well, as it can be seen in Table 2, with Linear Regression following in the results. Interestingly, unigrams seem to be the feature that boosts the classification performance more than all the other features we examine, while the highest performance is observed for the vectors excluding trigrams. Moreover, the binary field representing the existence of a hashtag in the tweet affects the results, as in all the experiments, the performance records smaller values without it. It can also be observed that all three algorithms perform better for positive and negative tweets than they do for neutral messages.

To further evaluate our system, we conducted a user study in which results from our approach were compared to those from user. The online survey using

Table 4. Ternary Classification - F-Measure for Logistic Regression

Features	Positive	Negative	Neutral	Total
Complete feature vector	0.628	0.592	0.542	0.591
w/o Unigrams	0.596	0.457	0.451	0.51
w/o Bigrams	0.616	0.6	0.546	0.59
w/o Trigrams	0.649	0.623	0.572	0.618
w/o User	0.625	0.6	0.54	0.592
w/o Hashtag	0.612	0.591	0.526	0.58
w/o URL	0.613	0.598	0.537	0.585
w/o POS Tags	0.646	0.585	0.512	0.587

Table 5. Ternary Classification - F-Measure for Decision Trees

Features	Positive	Negative	Neutral	Total
Complete feature vector	0.646	0.727	0.557	0.643
w/o Unigrams	0.57	0.681	0.549	0.597
w/o Bigrams	0.647	0.729	0.557	0.644
w/o Trigrams	0.646	0.728	0.557	0.644
w/o User	0.646	0.727	0.557	0.643
w/o Hashtag	0.639	0.601	0.529	0.594
w/o URL	0.64	0.615	0.554	0.606
w/o POS Tags	0.659	0.729	0.56	0.65

Ruby on Rails[5] contained 220 tweets of the test set of the dataset used for the ternary classification. 10 students associated with the University of Patras manually classified the tweets, and in following we compared the classification results of the best classifier to the users' responses. As the corresponding classifier, we choose Naive Bayes without Trigrams as it achieves the best F-Measure for all sentiments in ternary classification. Moreover, for each tweet, we selected the sentiment that appears the most in students' selections.

The percentages of corrected classified tweets are presented in following Fig. 2. We can observe that our proposed algorithm seems to achieve notable accuracy when dealing with neutral tweets, whereas positive and sentiment tweets do not have accurate precision. One possible explanation is the fact that the majority of the specific dataset contains tweets that are classified as neutral.

---

[5] http://sentipoll.herokuapp.com/.

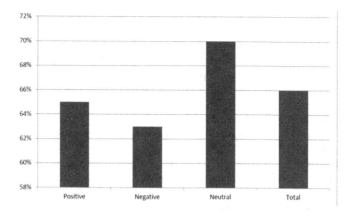

Fig. 2. Percentages of corrected classified tweets from Naive Bayes

7   Conclusions and Future Work

In our work, we have presented a tool that analyzes microblogging messages regarding their sentiment using machine learning techniques. More specifically, two datasets are utilized; a big dataset of tweets classified as positive or negative (binary), and a smaller one that consists of tweets classified as positive, negative or neutral (ternary). On the binary case, we examine the influence of the size of the dataset in relation with the performance of the sentiment analysis algorithms, while on the ternary case, we measure the system's accuracy regarding the different features extracted from the input. All the classification algorithms are implemented in Apache Spark cloud framework using the Apache Spark's Machine Learning library, entitled MLlib. Moreover, a user study was also conducted where University students manually classified tweets with the aim of validating our proposed tool accuracy.

As future work, we plan to further investigate the effect of different features on the input vector as well as utilize bigger datasets. Furthermore, we aim at experimenting with different clusters and evaluate Spark's performance in regards to time and scalability. Moreover, we plan on creating an online service that takes advantage of Spark Streaming, which is an Apache Spark's library for manipulating streams of data that provides users with real time analytics about sentiments of requested topics. Ultimately, personalization methods may be used to enhance the system's performance.

References

1. Agarwal, A., et al.: Sentiment analysis of Twitter data. In: Workshop on Languages in Social Media (2011)
2. Boiy, E., Moens, M.-F.: A machine learning approach to sentiment analysis in multilingual web texts. Inf. Retrieval **12**(5), 526–558 (2008)

3. Bollen, J., Mao, H., Pepe, A.: Twitter sentiment and socio-economic phenomena. In: International Conference on Web and Social Media (ICWSM) (2011)
4. Chikersal, P., Poria, S., Cambria, E.: Sentiment analysis of tweets by combining a rule-based classifier with supervised learning. In: International Workshop on Semantic Evaluation (SemEval), pp. 647–651 (2015)
5. Chinthala, S., et al.: Sentiment analysis on twitter streaming data. in: Emerging ICT for Bridging the Future-Proceedings of the Annual Convention of the Computer Society of India (CSI), vol. 1 (2015)
6. Dean, J., Ghemawat, S.: MapReduce: simplified data processing on large clusters. Commun. ACM **51**(1), 107–113 (2008)
7. Go, A., Bhayani, R., Huang, L.: Twitter sentiment classification using distant supervision. CS224N Project Report, Stanford 1, vol. 12 (2009)
8. Hodeghatta, U.R.: Sentiment analysis of hollywood movies on Twitter. In: IEEE/ACM International Conference on Advances in Social Networks Analysis and Mining (ASONAM), pp. 1401–1404 (2013)
9. Kanavos, A., Perikos, I., Vikatos, P., Hatzilygeroudis, I., Makris, C., Tsakalidis, A.: Modeling ReTweet diffusion using emotional content. In: Artificial Intelligence Applications and Innovations (AIAI), pp. 101–110 (2014)
10. Kanavos, A., Perikos, I., Vikatos, P., Hatzilygeroudis, I., Makris, C., Tsakalidis, A.: Conversation emotional modeling in social networks. In: IEEE International Conference on Tools with Artificial Intelligence (ICTAI), pp. 478–484 (2014)
11. Kanavos, A., Perikos, I.: Towards detecting emotional communities in Twitter. In: IEEE International Conference on Research Challenges in Information Science (RCIS), pp. 524–525 (2015)
12. Kanavos, A., Perikos, I., Hatzilygeroudis, I., Tsakalidis, A.: Integrating user's emotional behavior for community detection in social networks. In: International Conference on Web Information Systems and Technologies (WEBIST) (2016)
13. Kim, S.M., Hovy, E.: Determining the sentiment of opinions. In: International Conference on Computational Linguistics, p. 1367 (2004)
14. Naveed, N., Gottron, T., Kunegis, J., Alhadi, A.C.: Bad news travel fast: a content-based analysis of interestingness on Twitter. Web Science, Article No. 8 (2011)
15. Nodarakis, N., Sioutas, S., Tsakalidis, A., Tzimas, G.: Large scale sentiment analysis on Twitter with Spark. In: EDBT/ICDT Workshops (2016)
16. Pak, A., Paroubek, P.: Twitter as a corpus for sentiment analysis and opinion mining. In: LREC, vol. 10 (2010)
17. Pang, B., Lee, L.: Opinion mining and sentiment analysis. Found. Trends Inf. Retrieval **2**(1–2), 1–135 (2008)
18. Pang, B., Lee, L., Vaithyanathan, S.: Thumbs up? Sentiment classification using machine learning techniques. In: ACL Conference on Empirical methods in Natural Language Processing, pp. 79–86 (2002)
19. Poonam, W.: Twitter sentiment analysis with emoticons. Int. J. Eng. Comput. Sci. **4**(4), 11315–11321 (2015)
20. Suttles, J., Ide, N.: Distant supervision for emotion classification with discrete binary values. In: CICLing, pp. 121–136 (2013)
21. Toutanova, K., Klein, D., Manning, C., Singer, Y.: Feature-rich part-of-speech tagging with a cyclic dependency network. In: HLT-NAACL, pp. 252–259 (2003)
22. Turney, P.D.: Semantic orientation applied to unsupervised classification of reviews. In: Annual Meeting on Association for Computational Linguistics, pp. 417–424 (2002)

23. Wang, H., Can, D., Kazemzadeh, A., Bar, F., Narayanan, S.: A system for real-time twitter sentiment analysis of 2012 us presidential election cycle. In: ACL System Demonstrations, pp. 115–120 (2012)
24. Wilson, T., Wiebe, J., Hoffmann, P.: Recognizing contextual polarity in phrase-level sentiment analysis. In: Conference on Human Language Technology and Empirical Methods in Natural Language Processing, pp. 347–354 (2005)
25. Yamamoto, Y., Kumamoto, T., Nadamoto, A.: Role of emoticons for multidimensional sentiment analysis of Twitter. In: International Conference on Information Integration and Web-based Applications Services (iiWAS), pp. 107–115 (2014)

(A)kNN Query Processing on the Cloud: A Survey

Nikolaos Nodarakis[1]([✉]), Angeliki Rapti[1], Spyros Sioutas[2],
Athanasios K. Tsakalidis[1], Dimitrios Tsolis[3], Giannis Tzimas[4],
and Yannis Panagis[5]

[1] Computer Engineering and Informatics Department,
University of Patras, 26504 Patras, Greece
{nodarakis,arapti,tsak}@ceid.upatras.gr
[2] Department of Informatics, Ionian University, 49100 Corfu, Greece
sioutas@ionio.gr
[3] Department of Cultural Heritage, Management and New Technologies,
University of Patras, 26504 Patras, Greece
dtsolis@upatras.gr
[4] Computer and Informatics Engineering Department, Technological Educational,
Institute of Western Greece, 26334 Patras, Greece
tzimas@cti.gr
[5] Centre of Excellence for International Courts, University of Copenhagen,
1455 Copenhagen, Denmark
ioannis.panagis@jur.ku.dk

Abstract. A k-nearest neighbor (kNN) query determines the k nearest points, using distance metrics, from a given location. An all k-nearest neighbor (AkNN) query constitutes a variation of a kNN query and retrieves the k nearest points for each point inside a database. Their main usage resonates in spatial databases and they consist the backbone of many location-based applications and not only. Although (A)kNN is a fundamental query type, it is computationally very expensive. During the last years a multiplicity of research papers has focused around the distributed (A)kNN query processing on the cloud. This work constitutes a survey of research efforts towards this direction. The main contribution of this work is an up-to-date review of the latest (A)kNN query processing approaches. Finally, we discuss various research challenges and directions of further research around this domain.

Keywords: Big data · Nearest neighbor · MapReduce · NoSQL · Query processing

1 Introduction

A k-nearest neighbor query [46] computes the k nearest points, using distance metrics, from a specific location and is an operation that is widely used in spatial databases. An all k-nearest neighbor query constitutes a variation of a kNN query

© Springer International Publishing AG 2017
T. Sellis and K. Oikonomou (Eds.): ALGOCLOUD 2016, LNCS 10230, pp. 26–40, 2017.
DOI: 10.1007/978-3-319-57045-7_3

and retrieves the k nearest points for each point inside a dataset in a single query process. There is a wide diversity of applications that (A)kNN queries can be harnessed, such as the classification problem. Furthermore, they are widely used by location based services [24]. For example, consider users that send their location to a web server to process a request using a position anonymization system in order to protect their privacy from insidious acts. This anonymization system may use a kNN algorithm to calculate the k nearest neighbors for a specific user. After that, it sends to the server the locations of the neighbors along with the location of the user that made the request at the first place. In addition, many algorithms have been developed to optimize and speed up the join process in databases using the kNN approach.

Although (A)kNN is a fundamental query type, it is computationally very expensive. The naive approach is to search for every point the whole dataset in order to estimate its k-NN list. This leads to an $O\left(n^2\right)$ time complexity assuming that n is the cardinality of the dataset. As a result, quite a few centralized algorithms and structures (M-trees, R-trees, space-filling curves, etc.) have been developed towards this direction [11,16,23,25,66]. However, as the volume of datasets grows rapidly even these algorithms cannot cope with the computational burden produced by such queries. Consequently, high scalable implementations are required. Cloud computing technologies provide tools and infrastructure to create such solutions and manage the input data in a distributed way among multiple servers. The most popular and notably efficient tool is the *MapReduce* [17] programming model, developed by Google, for efficiently processing large-scale data.

In this spirit, we perform a thorough survey of existing distributed (A)kNN methods regarding only the cloud infrastructure (e.g. MapReduce model, NoSQL databases, etc.). We avoid studying other environments for such solutions (e.g. OpenMP/MPI, GPU-enabled solutions) since this excessively broadens the range of existing applications. Plus, the concept is a bit different in this case (e.g. cloud methods support fault tolerant behavior in contrary to an OpenMP/MPI approach). We leave the study of such techniques for future work, as an extension of the current paper. Some efforts to create an overview of (A)kNN techniques have been made in the past [1,10,19,33], but none of them refer to distributed solutions. Moreover, there exist a few surveys that are complementary to our work [47,48], but focus solely to kNN join methods on MapReduce. On the other hand, we elaborate on cloud-based (A)kNN methods in a much broader fashion regardless the type of application, thus providing an holistic and up-to-date overview on this research domain. The purpose of this paper is to concentrate all existing literature around this domain in order to make easy for other researchers to find the best solution (or deploy their own) tailored to their needs.

The rest of the paper is structured as follows: in Sect. 2 we present the taxonomy and classification table. In Sect. 3, we delve into more details for each classification dimension of our taxonomy and we outline the main findings, while in Sect. 4 we conclude the paper and propose some future research directions.

2   Taxonomy and Classification

In order to provide a concise classification of the existing approaches to cloud
elasticity, we first propose a taxonomy that will enable our work to shed light on
the differentiating aspects of the various proposals. The taxonomy is summarized
in Fig. 1 and consists of the following dimensions:

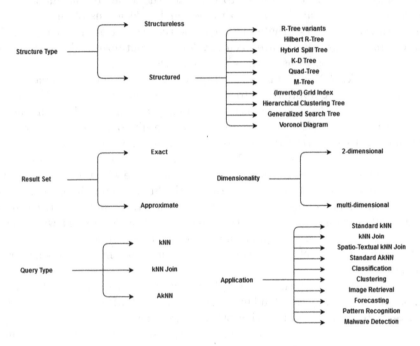

Fig. 1. Classification scheme

2.1   Structure Type

This aspect is divided into two categories: (1) the structureless solutions and (2)
the structured solutions. The former indicates a batch-based approach applied
on the entire volume of data at once. It is suitable for AkNN queries, since it
is faster than building an index and accessing it for an enormous test set. Its
disadvantage lies in the fact that if the data are amended, the whole process
needs to be repeated. The structured solutions rely on data preprocessing, in
order to build an efficient index to answer the queries quite fast. Thus, it is
usually appropriate for kNN queries. The up-to-date techniques build distributed versions of well-known structures, like R-Tree (and its variants), Hilbert
R-Tree, Hybrid Spill Tree, K-D Tree, Quad-Tree, M-Tree, Inverted Grid Index,
Hierarchical Clustering Tree, Generalized Search Tree, Voronoi Diagram, etc.
The nodes of the cluster may form a global index or each of them may contain a

local index. In the first case, only the appropriate nodes contribute in an index operation (e.g. search), thus sparing resources, while in the latter case all nodes participate in the executed operation.

2.2 Result Set

As mentioned and before, (A)kNN queries are computationally intensive, especially on a massive scale as in the cloud era. As a result, the researchers have investigated many alternatives in order to minimize the execution cost of an (A)kNN query. This has led to the development of approximate methods that return a result set that is correct with high probability. These methods may use probabilistic models or heuristics and usually are orders of magnitude faster than exact solutions, however it is possible to fail under certain circumstances. Depending on the application and the necessity for a correct result, we can either apply approximate or exact techniques to process the given queries.

2.3 Dimensionality

A very significant perspective when dealing with (A)kNN queries is the dimensionality of the provided dataset. For some problems, such as location-based services, implementing methods that handle only 2-dimensional points is adequate. On the other hand, some techniques are obligated to process points for an arbitrary number of dimensions (e.g. classification problems). When designing a method that works only in 2-dimensions, we can make specific assumptions and implementation decisions in order to optimize it for the purpose it serves. However, it is quite difficult to generalize this solution for more dimensions if it is required in the future. On the other hand, generalized solutions are more stable if we achieve a concrete implementation, but often suffer from the curse of dimensionality especially when the provided input follows a power-law or zipfian distribution. The dimensionality perspective is an extremely important issue and must be taken under serious consideration when designing an approach for such query types.

2.4 Query Type

All methodologies presented in this work use variations of kNN queries that are classified in three categories. The most common type is the standard kNN query, where we investigate for the k-nearest neighbors of a single element. The rest of query types are slightly different and are often used describing the same task. The kNN join is a combination of the k-nearest neighbor (kNN) query and the join operation and merges each point in a dataset R with its k-nearest neighbors in a dataset S. The AkNN query can alternatively be viewed as a kNN self-join query or a special-case kNN join where R and S are identical. For the rest of the paper, we assume that kNN join and AkNN are two distinct query types.

2.5   Application

Finally, this classification category refers to the applicability of the techniques that are introduced in the context of this survey. It is well known that kNN queries are fundamental and consist the backbone of many applications in data mining, machine learning, etc. The application type of existing solutions is distinguished into the following categories: (1) Standard kNN, (2) kNN Join, (3) Spatio-textual kNN Join, (4) Standard AkNN, (5) Classification, (6) Clustering, (7) Image Retrieval, (8) Forecasting, (9) Pattern Recognition and (10) Malware Detection.

2.6   Classification Table

Based on the taxonomy above, we classify the existing proposals for distributed (A)kNN query processing on the cloud as shown in Table 1. Each citation studied in this survey is represented by a row in the table, and each cell of the row denotes the class it belongs for each distinct category that is displayed in Fig. 1. In case of Structure Type aspect, we use a check-mark or a cross-mark to declare whether the method is structureless or not. Similarly, if the method is structured we write down the index it builds, else we use a cross-mark.

3   Overview of (A)kNN Methods

In this section, we provide details with respect to the main solution approaches for each taxonomy dimension in turn and focus on the key features of them.

3.1   Structure Type Taxonomy

This dimension indicates whether a solution builds a structure to process the query or not. As seen from Table 1, the majority of solutions rely on some kind of indexing scheme in order to prune as many as much redundant calculations, thus minimizing the required computational burden. These techniques often need a considerable amount of time to preprocess the input data and construct the index, but in the end they pay off since the subsequent queries are answered much faster than running a batch-based process for each query individually. For example, [22] uses indexes to perform kNN queries and its throughput is two orders of magnitude better than Hadoop, while the method presented in [31] further improve its performance. Index-based methods are more suitable when the input data undergo many alterations (i.e. update/insert/delete operations) since we do not need to build the index from scratch. Furthermore, the vast majority of approaches advocate only kNN queries, except from [3–5,20,21,53, 56,58,59,63] which support and kNN Join or only kNN Join queries.

A wide diversity of indexing schemes have been adapted to fit into the distributed nature of cloud paradigm. The R-Tree and its variants it the most commonly used structure since it is quite easy to implement it [3–5,12,13,20,22,29,36,38,56–59,62–64]. On the other hand, this kind of trees have a major deficiency which is

Table 1. The classification scheme based on our taxonomy

Citation	Structure Type		Result Set	Dimensionality	Query Type	Application
	Structureless	Structured				
[35]	✓	✗	Exact	Multi-dimensional	kNN Join	Classification
[52]	✓	✗	Exact	Multi-dimensional	AkNN	Forecasting
[21]	✗	Hilbert R-Tree	Exact	Multi-dimensional	kNN Join	kNN Join
[58]	✗	R-Tree	Exact	Multi-dimensional	kNN Join	kNN Join
[45]	✓	✗	Exact	Multi-dimensional	AkNN	Standard AkNN
[60]	✓	✗	Exact	2-dimensional	kNN Join	Spatio-Textual kNN Join
[30]	✗	Hybrid Spill Tree	Approximate	Multi-dimensional	kNN	Clustering
[65]	✗	Grid Index	Exact	2-dimensional	AkNN	Standard AkNN
[2]	✗	K-D-Tree	Exact	Multi-dimensional	kNN	Standard kNN
[44]	✗	Any*	Exact	Multi-dimensional	kNN	Pattern Recognition
[15]	✓	✗	Exact	Multi-dimensional	AkNN	Standard AkNN
[7]	✗	K-D-Tree	Exact	Multi-dimensional	kNN	Image Retrieval
[18]	✓	✗	Exact	Multi-dimensional	AkNN	Classification
[40]	✓	✗	Exact	Multi-dimensional	AkNN	Classification
[63]	✗	R-Tree	Exact/Approximate	Multi-dimensional	kNN Join	kNN Join
[32]	✓	✗	Exact	Multi-dimensional	kNN Join	kNN Join
[13]	✗	R-Tree	Exact	Multi-dimensional	kNN	Standard kNN
[14]	✓	✗	Approximate	Multi-dimensional	kNN Join	Malware Detection
[62]	✗	R-Tree/Quad-Tree	Exact	2-dimensional	kNN	Standard kNN
[38]	✗	Grid Index/R(+)-Tree	Exact	2-dimensional	kNN	Standard kNN
[5]	✗	R*-Tree	Exact	Multi-dimensional	kNN/kNN Join	Standard kNN/kNN Join
[56]	✗	R*-Tree	Exact	Multi-dimensional	kNN/kNN Join	Standard kNN/kNN Join
[3]	✗	R*-Tree	Exact	Multi-dimensional	kNN/kNN Join	Standard kNN/kNN Join
[59]	✗	R*-Tree	Exact	Multi-dimensional	kNN Join	kNN Join
[57]	✗	R-Tree	Exact	Multi-dimensional	kNN	Standard kNN
[26]	✗	Inverted Grid Index	Exact	2-dimensional	kNN	Standard kNN
[27]	✗	Inverted Grid Index	Exact	2-dimensional	kNN	Standard kNN
[41]	✓	✗	Exact	Multi-dimensional	AkNN	Classification
[34]	✓	✗	Exact	Multi-dimensional	kNN Join	Classification
[42]	✓	✗	Exact	Multi-dimensional	kNN Join	Classification
[53]	✗	Quad-Tree	Exact	2-dimensional	kNN/kNN Join	Standard kNN/kNN Join
[39]	✗	Quad-Tree/K-D-Tree	Exact	Multi-dimensional	kNN	Standard kNN
[51]	✓	✗	Exact	2-dimensional	AkNN	Standard AkNN
[54]	✓	✗	Exact	Multi-dimensional	kNN Join	Classification
[43]	✓	✗	Exact	Multi-dimensional	kNN Join	Classification
[29]	✗	R-Tree	Exact	Multi-dimensional	kNN	Standard kNN
[28]	✗	M-Tree	Exact	2-dimensional	kNN	Standard kNN
[55]	✓	✗	Exact	Multi-dimensional	AkNN	Classification
[36]	✗	R(+)-Tree	Exact	Multi-dimensional	kNN	Standard kNN
[61]	✓	✗	Exact	2-dimensional	AkNN	Standard AkNN
[12]	✗	R-Tree/ K-D-Tree/Voronoi Diagram	Exact	Multi-dimensional	kNN	Standard kNN
[49]	✓	✗	Approximate	Multi-dimensional	kNN	Standard kNN
[64]	✗	Grid Index/ R-Tree/Quad-Tree	Exact	2-dimensional	kNN	Standard kNN
[37]	✗	Hierarchical Clustering Tree	Approximate	Multi-dimensional	kNN	Standard kNN
[31]	✗	Generalized Search Tree	Exact	Multi-dimensional	kNN	Standard kNN
[8]	✗	Quad-Tree	Exact	Multi-dimensional	kNN	Standard kNN
[20]	✗	R-Tree/K-D-Tree	Exact	Multi-dimensional	kNN/kNN Join	Standard kNN/kNN Join
[9]	✓	✗	Exact	Multi-dimensional	kNN/kNN Join	Standard kNN/kNN Join
[67]	✓	✗	Exact	Multi-dimensional	kNN/AkNN	Standard kNN/AkNN
[22]	✗	Grid Index/R(+)-Tree	Exact	Multi-dimensional	kNN	Standard kNN
[50]	✓	✗	Exact	Multi-dimensional	AkNN	Pattern Recognition
[4]	✗	R*-Tree/Voronoi Diagram	Exact	Multi-dimensional	kNN/kNN Join	Standard kNN/kNN Join
[68]	✗	Quad-Tree	Exact	Multi-dimensional	kNN	Standard kNN
[6]	✗	Voronoi-Diagram	Exact	2-dimensional	kNN	Standard kNN

* ⟶ Any data structure that can answer kNN queries

the overlapping between the space covered in internal nodes. As a result, the query response time can be bigger compared to other data structures. Consequently, quite a few other indexing schemes have been adapted to needs of cloud computing, such as K-D-Tree [2,7,12,20,39], Quad-Tree [8,39,53,62,64,68], Grid Index [22,26,27,38,64,65], Voronoi Diagram [4,6,12], etc.

An index can either be local or global. The former means that each node of the cluster maintains an indexing structure. So, when a query arrives we search simultaneously in all nodes and the partial results are merged into the final answer. This indexing scheme is much simpler than the global index scheme, but its main disadvantage lies in the fact that it is required to visit all nodes to get the answer, even if it is not necessary. On the other hand, when building a global index each node of the cluster contains a part of it. Thus, we search only on the nodes that actually contain the term we are looking for and do not stress with extra burden the rest of them. The deficiency of the latter, is that the construction of a global index is a much more complex task.

When dealing with (A)kNN queries, index structures are not always the cure. More specifically, when we need to run batch-based processes for a massive amount of elements we usually prefer structureless solutions. Another reason to use such approaches is the static nature of the data (e.g. when we process historic data). A structureless solution is often much faster than an index-based, since it does not waste any time to preprocess the data and construct the index. The vast majority of methods that do not use an indexing scheme support AkNN and kNN join queries which are computationally intensive [9,14,15,18,32,34,35,40–43,45,50–52,54,55,60,61,67]. The only exception is the work presented in [49] which employs a LSH scheme to calculate kNN queries, while in [9,67] simple kNN queries are supported along with other query types.

3.2 Result Set Taxonomy

In this aspect we examine whether the outcome provided by each method is exact or approximate. The overwhelming majority of approaches employ techniques that return the exact set a query. For a great variety of applications, such as location-based services (LBS), it is very crucial to obtain the exact kNN set because they rely on high accuracy performance [6,26–28,38,51,53,60–62,64,65]. Other examples with similar requirements for precise solutions are the pattern recognition paradigm which is discussed in [44,50], the image retrieval problem [7] and the forecasting problem [52]. For other problem types (e.g. classification) either approximate or exact approaches have been proposed. For example, in the context of [47,48,63] both exact and approximate techniques for processing kNN joins are studied and compared. Besides the accuracy of the outcome, exact and approximate methods differ significantly on the aspect of execution time with the former to be much slower than the latter.

An approximate method has been designed to provide a very good estimation of the exact answer to a query with high probability. They rely on probabilistic models or heuristics and can notably decrease the running time of a problem solution. They are often designed for NP-hard problems where exact solutions need enormous resources and running time and, consequently, are impractical to use. For instance, the authors in [49] discuss a distributed locality sensitive hashing scheme (LSH) to process kNN queries on top of MapReduce while in [14] a malware detection method is displayed using similarity joins. Another suitable domain for approximate kNN query processing are the clustering problems. In [30] a method is presented to cluster billions of images using a Hybrid Spill Tree to advocate approximate kNN queries. Furthermore, in [37] a Hierarchical Clustering Tree is proposed to efficiently process approximate kNN queries.

3.3 Dimensionality Taxonomy

Dimensionality plays an important role to the outline of a solution in any kind of problem. We can distinguish two main concepts. In the first one, the design of the approach is dimension-specific while in the second one the method is dimension-independent. A method that follows the former approach exploits techniques that are applied the to predefined dimension cardinality in order to be optimal. However, in case of scale up in the number of dimensions, the solution either ceases to work or its performance seriously degrades due to the curse of dimensionality. A method that follows the latter approach can theoretically work efficiently for any number of dimensions. In practise, according to the technique and the domain it is applied, there is an upper threshold on this number before the curse of dimensionality comes to the foreground (e.g. [40,41]). The only drawback is that they cannot take advantage of dimension-specific techniques that can boost their performance. Dimension-independent approaches are more preferable when the number of dimensions is arbitrary or unknown, while dimension-specific approaches are useful for applications that apply for a small number of dimensions which is known.

The papers presented in this survey adapt either multi-dimensional (this is the dominant category) or 2-dimensional techniques. With a first glance at Table 1 we observe that all 2-dimensional solutions apply to LBS and GIS (Geographical Information Systems) applications. The reason is that input data represent points with longitude and latitude on a map or a planar layer. The authors in [53,62] propose state-of-the-art index-assisted in memory spatial data processing on top of Spark while in [38,65] two GIS frameworks for spatial queries on top of Hadoop are studied. Meanwhile, the work in [51] performs a comparative analysis for GIS data processing between Hadoop and Spark implementations and investigates their advantages and disadvantages. A very popular technique for processing kNN queries on the plane is the space decomposition using a grid index [26,27,61] or a voronoi diagram [6]. But, the most widely used structure for answering kNN queries is the M-Tree and a distributed version of it (NSM-Tree) is proposed in [28]. A different perspective of kNN queries on a plane is described in detail in [64] and the performance of a Storm-based framework on real-time queries is evaluated. Finally, the authors in [60] present a spatio-textual kNN join method on sensor networks.

On the other hand, the applicability of multi-dimensional applications spans in a wide domain area. According to Table 1 the most common problems that handles a multi-dimensional solution is the kNN join (e.g. [3,5,21,58,59]), the AkNN query (e.g. [15,45,61]) and the classification to a set of categories (e.g. [18,34,54]). This makes sense since the vector space of each element can be of arbitrary size in these problem categories. More precisely, the number of dimensions may vary from a single digit number [40,41] to a few tens [32,63] or thousands [30,42,43,55]. Moreover, the multi-dimensional proposed techniques can also be approximate compared to the respective 2-dimensional which deliver only exact results. From the above, we conclude that multi-dimensional solutions are much more diverse than 2-dimensional approaches.

3.4 Query Type Taxonomy

In this standpoint, we examine the partition of the proposed solutions between the different query types. A more detailed look at Table 1 reveals some correlations between the structure type and the query type. Firstly, the kNN query type is mainly supported by structured solutions. The only 2 exceptions are the works presented in [49,67]. In the first paper, the authors estimate the query results using a distributed LSH scheme, while in the second one the authors discuss a high level implementation of classic kNN ($O(n^2)$ time complexity) in the MapReduce model. The applications that usually need simple kNN queries process dynamic data, which means multiple amendments in the original dataset (i.e. insert/update/delete). Consequently, an indexing scheme is the most suitable for this kind of data operations.

On the other hand, exactly the opposite happens in the case of AkNN query. Excluding the framework described in [65], all other methods follow a structureless strategy. This type of query is computationally intensive and it is common to be executed on static data, since it is executed as a batch-based process. So, it is more preferable to deploy an efficient one-pass method that is applied directly to the input data, rather than building an indexing structure. The construction of the index requires a notable amount of time without any counterbalance, since the data do not change and we can not exploit any low cost operations (e.g. insert new elements).

Considering the kNN join query, the number of approaches balances between structured and structureless strategies. The structureless implementations rely on some kind of space partitioning techniques, such as block partitioning [34,60], voronoi diagram partitioning [32], etc. A similar approach is followed and in structured solutions, but the difference is that now the partitioning is applied according to the position of an element in the structure (e.g. [53,59]).

3.5 Application Taxonomy

The application range of the studied literature is quite extensive. As it is expected, most of the proposed solutions are applicable for the standard kNN operation (e.g. [2,3,5,8,9,13]). This operation is quite generic and can be further refined to more specific applications, however the authors of these methods do not delve into more details. The advantage of these implementations is that they may (theoretically) be applied to any problem category, with the drawback that they can not exploit any domain-specific techniques to boost the query performance. Similarly, the same things apply for the standard AkNN [15,45,51,61,65,67] and kNN join operation [3–5,20,21,32,53,56,58,59,63,67]. In any case, these query types are mainly used in LBSs. For example, consider users that send their location to a web server to process a request using a position anonymization system in order to protect their privacy from insidious acts. This anonymization system may use an AkNN algorithm to calculate the k nearest neighbors for each user. After that, it sends to the server the locations of the neighbors along with the location of the user that made the request at the first place.

Another accustomed application of (A)kNN queries is the classification domain. In [18] the authors propose an algorithm where they first conduct a k-means clustering to separate the whole dataset into several parts, each of which is then conducted kNN classification. A simple block-based partition approach is implemented in Hadoop and Spark in [34,35], respectively. A classification method in Hadoop, that can be applied on multi-dimensional points in LBSs, is presented in [40,41]. The authors exploit space decomposition techniques to achieve the desired outcome and by far optimize the method presented in [61]. Moreover, the classification paradigm is extensively used when performing sentiment analysis tasks, as in [42,43] where a classification procedure of diverse sentiment types on Twitter data is performed. The kNN classification scheme has also been utilized in prototype reduction in big data [54] and in duplicate detection in adverse drug reaction databases [55].

The remaining subset in literature related to (A)kNN queries involves less popular application types. For instance, only one research effort has been made towards the clustering problem [30]. The research community has also drawn attention to the pattern recognition domain. In [44] a detailed description of an algorithm for building the kNN graph for large high-dimensional point sets is given. In addition, a method for recognizing, tagging and classifying facial images has been deployed in [50]. Similarly, a kNN approach (based on K-D-Trees) for image retrieval is proposed in [7]. A different point of view of kNN queries is displayed in [52], where a nearest neighbours-based algorithm for big time series data forecasting is evaluated. Another interesting perspective is presented in [14], where kNN is used for malware detection on HTTPS data. Finally, a similar problem to kNN join on sensor networks is discussed in [60].

4   Conclusions and Future Steps

To conclude, in the context of this survey we investigated in detail existing literature that deals with (A)kNN queries on cloud environments. We divided the presented papers based on a set of taxonomies and classified each proposed method according to the category it belongs in each taxonomy (Table 1). For each aspect in all taxonomies we presented its main features, as long as its advantages and disadvantages.

Although a big set of solutions that answer (A)kNN query on the cloud exists, a few things can be further researched. One interesting perspective is the study of (A)kNN methods on different distributed frameworks, like OpenMP/MPI or GPU-based solutions, and their comparison with cloud solutions. Finally, another possible research direction is the overview of techniques that process kNN-like queries, for instance RNN (Reverse Nearest Neighbor) or top-k queries.

References

1. Abbasifard, M.R., Ghahremani, B., Naderi, H.: A survey on nearest neighbor search methods. Int. J. Comput. Appl. **95**, 39–52 (2014)
2. Abdelsadek, A., Hefeeda, M.: DIMO: distributed index for matching multimedia objects using MapReduce. In: Proceedings of the 5th ACM Multimedia Systems Conference, pp. 115–126. ACM, New York (2014)
3. Aji, A., Wang, F.: High performance spatial query processing for large scale scientific data. In: Proceedings of the on SIGMOD/PODS 2012 PhD Symposium, pp. 9–14. ACM, New York (2012)
4. Aji, A., Wang, F., Saltz, J.H.: Towards building a high performance spatial query system for large scale medical imaging data. In: Proceedings of the 20th International Conference on Advances in Geographic Information Systems, pp. 309–318. ACM, New York (2012)
5. Aji, A., Wang, F., Vo, H., Lee, R., Liu, Q., Zhang, X., Saltz, J.: Hadoop GIS: a high performance spatial data warehousing system over MapReduce. Proc. VLDB Endow. **6**, 1009–1020 (2013)
6. Akdogan, A., Demiryurek, U., Kashani, F.B., Shahabi, C.: Voronoi-based geospatial query processing with MapReduce. In: Proceedings of the IEEE 2nd International Conference on Cloud Computing Technology and Science, pp. 9–16. IEEE Computer Society, Washington, DC (2010)
7. Aly, M., Munich, M., Perona, P.: Distributed Kd-trees for retrieval from very large image collections. In: Proceedings of the British Machine Vision Conference (BMVC) (2011)
8. Andreica, M.I., Tapus, N.: Sequential and MapReduce-based algorithms for constructing an in-place multidimensional quad-tree index for answering fixed-radius nearest neighbor queries. Acta Universitatis Apulensis - Mathematics-Informatics, pp. 131–151 (2012)
9. Baig, F., Mehrotra, M., Vo, H., Wang, F., Saltz, J., Kurc, T.: SparkGIS: efficient comparison and evaluation of algorithm results in tissue image analysis studies. In: Wang, F., Luo, G., Weng, C., Khan, A., Mitra, P., Yu, C. (eds.) Big-O(Q)/DMAH -2015. LNCS, vol. 9579, pp. 134–146. Springer, Cham (2016). doi:10.1007/978-3-319-41576-5_10
10. Bhatia, N.: Vandana: Survey of Nearest Neighbor Techniques. CoRR abs/1007.0085 (2010)
11. Böhm, C., Krebs, F.: The k-nearest neighbour join: turbo charging the KDD process. Knowl. Inf. Syst. **6**, 728–749 (2004)
12. Candan, K.S., Nagarkar, P., Nagendra, M., Yu, R.: RanKloud: a scalable ranked query processing framework on hadoop. In: Proceedings of the 14th International Conference on Extending Database Technology, pp. 574–577. ACM, New York (2011)

13. Cary, A., Sun, Z., Hristidis, V., Rishe, N.: Experiences on processing spatial data with MapReduce. In: Winslett, M. (ed.) SSDBM 2009. LNCS, vol. 5566, pp. 302–319. Springer, Heidelberg (2009). doi:10.1007/978-3-642-02279-1_24

14. Cech, P., Kohout, J., Lokoc, J., Komárek, T., Marousek, J., Pevný, T.: Feature extraction and malware detection on large HTTPS data using MapReduce. In: Amsaleg, L., Houle, M.E., Schubert, E. (eds.) SISAP 2016. LNCS, vol. 9939, pp. 311–324. Springer, Cham (2016). doi:10.1007/978-3-319-46759-7_24

15. Chatzimilioudis, G., Costa, C., Zeinalipour-Yazti, D., Lee, W.-C., Pitoura, E.: Distributed in-memory processing of all k nearest neighbor queries. IEEE Trans. Knowl. Data Eng. **28**, 925–938 (2016)

16. Chen, Y., Patel, J.M.: Efficient evaluation of all-nearest-neighbor queries. In: Proceedings of the 23rd IEEE International Conference on Data Engineering, pp. 1056–1065. IEEE Computer Society, Washington, DC (2007)

17. Dean, J., Ghemawat, S.: MapReduce: simplified data processing on large clusters. In: Proceedings of the 6th Symposium on Operating Systems Design and Implementation, pp. 137–150. USENIX Association, Berkeley (2004)

18. Deng, Z., Zhu, X., Cheng, D., Zong, M., Zhang, S.: Efficient kNN classification algorithm for big data. Neurocomputing **195**, 143–148 (2016)

19. Dhanabal, S., Chandramathi, S.: A review of various k-nearest neighbor query processing techniques. Int. J. Comput. Appl. **31**, 14–22 (2011)

20. Dong, X., Feifei, L., Bin, Y., Gefei, L., Liang, Z., Minyi, G.: Simba: efficient in-memory spatial analytics. In: Proceedings of the 2016 International Conference on Management of Data, pp. 1071–1085. ACM, New York (2016)

21. Du, Q., Li, X.: A novel KNN join algorithms based on Hilbert R-tree in MapReduce. In: Proceedings of the 3rd International Conference on Computer Science and Network Technology, pp. 417–420. IEEE (2013)

22. Eldawy, A., Mokbel, M.F.: SpatialHadoop: a MapReduce framework for spatial data. In: Proceedings of the 31st IEEE International Conference on Data Engineering, pp. 1352–1363. IEEE Computer Society, Washington, DC (2015)

23. Emrich, T., Graf, F., Kriegel, H.-P., Schubert, M., Thoma, M.: Optimizing all-nearest-neighbor queries with trigonometric pruning. In: Gertz, M., Ludäscher, B. (eds.) SSDBM 2010. LNCS, vol. 6187, pp. 501–518. Springer, Heidelberg (2010). doi:10.1007/978-3-642-13818-8_35

24. Gkoulalas-Divanis, A., Verykios, V.S., Bozanis, P.: A network aware privacy model for online requests in trajectory data. Data Knowl. Eng. **68**, 431–452 (2009)

25. Ioup, E., Shaw, K., Sample, J., Abdelguerfi, M.: Efficient AKNN spatial network queries using the M-Tree. In: Proceedings of the 15th Annual ACM International Symposium on Advances in Geographic Information Systems, pp. 46:1–46:4. ACM, New York (2007)

26. Ji, C., Dong, T., Li, Y., Shen, Y., Li, K., Qiu, W., Qu, W., Guo, M.: Inverted grid-based kNN query processing with MapReduce. In: Proceedings of the 7th ChinaGrid Annual Conference, pp. 25–32 (2012)

27. Ji, C., Li, Z., Qu, W., Xu, Y., Li, Y.: Scalable nearest neighbor query processing based on Inverted Grid Index. J. Network Comput. Appl. **44**, 172–182 (2014)

28. Kokotinis, I., Kendea, M., Nodarakis, N., Rapti, A., Sioutas, S., Tsakalidis, A.K., Tsolis, D., Panagis, Y.: NSM-Tree: efficient indexing on top of NoSQL databases. In: Post-proceedings of the 2nd International Workshop on Algorithmic Aspects of Cloud Computing (2016)

29. Liao, H., Jizhong, H., Jinyun, F.: Multi-dimensional index on hadoop distributed file system. In: Proceedings of the 2010 IEEE Fifth International Conference on Networking, Architecture, and Storage, pp. 240–249. IEEE Computer Society, Washington, DC, USA (2010)

30. Liu, T., Rosenberg, C., Rowley, H.A.: Clustering billions of images with large scale nearest neighbor search. In: Proceedings of the 8th IEEE Workshop on Applications of Computer Vision, p. 28. IEEE Computer Society (2007)

31. Lu, P., Chen, G., Ooi, B.C., Vo, H.T., Wu, S.: ScalaGiST: scalable generalized search trees for mapreduce systems [Innovative Systems Paper]. Proc. VLDB Endow. **7**, 1797–1808 (2014)

32. Lu, W., Shen, Y., Chen, S., Ooi, B.C.: Efficient processing of k nearest neighbor joins using mapreduce. Proc. VLDB Endow. **5**, 1016–1027 (2012)

33. Mahapatra, R.P., Chakraborty, P.S.: Comparative analysis of nearest neighbor query processing techniques. Procedia Comput. Sci. **57**, 1289–1298 (2015)

34. Maillo, J., Ramireza, S., Triguero, I., Herrera, F.: kNN-IS: an iterative spark-based design of the k-nearest neighbors classifier for big data. Knowledge-Based Systems (2016, in press)

35. Maillo, J., Triguero, I., Herrera, F.: A MapReduce-based k-nearest neighbor approach for big data classification. In: IEEE TrustCom/BigDataSE/ISPA, pp. 167–172. IEEE Computer Society, Washington, DC (2015)

36. Maleki, E.F., Azadani, M.N., Ghadiri, N.: Performance evaluation of spatialhadoop for big web mapping data. In: Proceedings of the 2016 Second International Conference on Web Research. IEEE Computer Society, Washington, DC (2016, to be published)

37. Muja, M., Lowe, D.G.: Scalable nearest neighbor algorithms for high dimensional data. IEEE Trans. Pattern Anal. Mach. Intell. **36**, 2227–2240 (2014)

38. Naami, K.M.A., Seker, S., Khan, L.: GISQF: an efficient spatial query processing system. In: Proceedings of the 2014 IEEE International Conference on Cloud Computing, pp. 681–688. IEEE Computer Society, Washington, DC (2014)

39. Nishimura, S., Das, S., Agrawal, D., Abbadi, A.E.: MD-HBase: a scalable multi-dimensional data infrastructure for location aware services. In: Proceedings of the 2011 IEEE 12th International Conference on Mobile Data Management, vol. 01, pp. 7–16. IEEE Computer Society, Washington, DC (2011)

40. Nodarakis, N., Pitoura, E., Sioutas, S., Tsakalidis, A., Tsoumakos, D., Tzimas, G.: Efficient multidimensional AkNN query processing in the cloud. In: Decker, H., Lhotská, L., Link, S., Spies, M., Wagner, R.R. (eds.) DEXA 2014. LNCS, vol. 8644, pp. 477–491. Springer, Cham (2014). doi:10.1007/978-3-319-10073-9_41

41. Nodarakis, N., Pitoura, E., Sioutas, S., Tsakalidis, A., Tsoumakos, D., Tzimas, G.: kdANN+: a rapid AkNN classifier for big data. Trans. Large-Scale Data Knowl. Centered Syst. **24**, 139–168 (2016)

42. Nodarakis, N., Sioutas, S., Tsakalidis, A., Tzimas, G.: Large scale sentiment analysis on Twitter with spark. In: Proceedings of the Workshops of the EDBT/ICDT 2016 Joint Conference, CEUR Workshop Proceedings, vol. 1558 (2016). CEUR-WS.org

43. Nodarakis, N., Sioutas, S., Tsakalidis, A., Tzimas, G.: MR-SAT: a MapReduce algorithm for big data sentiment analysis on Twitter. In: Proceedings of the 12th International Conference on Web Information Systems and Technologies, vol. 1, pp. 140–147. SciTePress (2016)

44. Plaku, E., Kavraki, L.E.: Distributed computation of the knn graph for large high-dimensional point sets. J. Parallel Distrib. Comput. **67**, 346–359 (2007)

45. Reyes-Ortiz, J.L., Oneto, L., Anguita, D.: Big data analytics in the cloud: spark on Hadoop vs MPI/OpenMP on Beowulf. Procedia Comput. Sci. **53**, 121–130 (2015)
46. Roussopoulos, N., Kelley, S., Vincent, F.: Nearest neighbor queries. In: Proceedings of the 1995 ACM SIGMOD International Conference on Management of Data, pp. 71–79. ACM, New York (1995)
47. Song, G., Rochas, J., Huet, F., Magoulès, F.: Solutions for processing K nearest neighbor joins for massive data on MapReduce. In: Proceedings of the 23rd Euromicro International Conference on Parallel, Distributed and Network-based Processing, March 2015, Turku, Finland (2015)
48. Song, G., Rochas, J., Huet, F., Magoulès, F.: K nearest neighbour joins for big data on MapReduce: a theoretical and experimental analysis. IEEE Trans. Knowl. Data Eng. **28**, 2376–2392 (2016)
49. Stupar, A., Michel, S., Schenkel, R.: RankReduce - processing K-nearest neighbor queries on top of MapReduce. In: Proceedings of the 8th Workshop on Large-Scale Distributed Systems for Information Retrieval, pp. 13–18. ACM, New York (2010)
50. Sun, K., Kang, H., Park, H.-H.: Tagging and classifying facial images in cloud environments based on KNN using MapReduce. Optik - Int. J. Light Electron Optics **126**, 3227–3233 (2015)
51. Sun, Z., Zhang, H., Liu, Z., Xu, C., Wang, L.: Migrating GIS big data computing from Hadoop to Spark: an exemplary study Using Twitter. In: Proceedings of the IEEE 9th International Conference on Cloud Computing, pp. 351–358. IEEE Computer Society, Washington, DC (2016)
52. Talavera-Llames, R.L., Pérez-Chacón, R., Martínez-Ballesteros, M., Troncoso, A., Martínez-Álvarez, F.: A nearest neighbours-based algorithm for big time series data forecasting. In: Martínez-Álvarez, F., Troncoso, A., Quintián, H., Corchado, E. (eds.) HAIS 2016. LNCS (LNAI), vol. 9648, pp. 174–185. Springer, Cham (2016). doi:10.1007/978-3-319-32034-2_15
53. Tang, M., Yu, Y., Malluhi, Q.M., Ouzzani, M., Aref, W.G.: LocationSpark: a distributed in-memory data management system for big spatial data. PVLDB **9**, 1565–1568 (2016)
54. Triguero, I., Peralta, D., Bacardit, J., García, S., Herrera, F.: MRPR: a MapReduce solution for prototype reduction in big data classification. Neurocomputing **150**(Part A), 331–345 (2015)
55. Wang, C., Karimi, S.: Parallel duplicate detection in adverse drug reaction databases with spark. In: Proceedings of the 19th International Conference on Extending Database Technology, pp. 551–562. ACM, New York (2016)
56. Wang, F., Aji, A., Vo, H.: High performance spatial queries for spatial big data: from medical imaging to GIS. SIGSPATIAL Special **6**, 11–18 (2014)
57. Wang, J., Wu, S., Gao, H., Li, J., Ooi, B.C.: Indexing multi-dimensional data in a cloud system. In: Proceedings of the 2010 ACM SIGMOD International Conference on Management of Data, pp. 591–602. ACM, New York (2010)
58. Wang, K., Han, J., Tu, B., Dai, J., Zhou, W., Song, X.: accelerating spatial data processing with MapReduce. In: Proceedings of the IEEE 16th International Conference on Parallel and Distributed Systems, pp. 229–236, IEEE Computer Society, Washington, DC (2010)
59. Xavier, P., Francis, F.S.: Improvisation to the R*-Tree kNN join principles in distributed environment. Int. J. Comput. Appl. **101**, 20–24 (2014)
60. Yang, M., Zheng, L., Lu, Y., Guo, M., Li, J.: Cloud-assisted spatio-textual k nearest neighbor joins in sensor networks. In: Proceedings of the 1st International Conference on Industrial Networks and Intelligent Systems, pp. 12–17. ICST, Gent, Belgium (2015)

61. Yokoyama, T., Ishikawa, Y., Suzuki, Y.: Processing all k-nearest neighbor queries in hadoop. In: Gao, H., Lim, L., Wang, W., Li, C., Chen, L. (eds.) WAIM 2012. LNCS, vol. 7418, pp. 346–351. Springer, Heidelberg (2012). doi:10.1007/978-3-642-32281-5_34

62. Yu, J., Wu, J., Sarwat, M.: GeoSpark: A cluster computing framework for processing large-scale spatial data. In: Proceedings of the 23rd International Conference on Advances in Geographic Information Systems, 03–06 November 2015. Association for Computing Machinery (2015)

63. Zhang, C., Li, F., Jestes, J.: Efficient parallel kNN joins for large data in MapReduce. In: Proceedings of the 15th International Conference on Extending Database Technology, pp. 38–49. ACM, New York (2012)

64. Zhang, F., Zheng, Y., Xu, D., Du, Z., Wang, Y., Liu, R., Ye, X.: Real-time spatial queries for moving objects using storm topology. ISPRS Int. J. Geo-Inf. **5**, 178 (2016)

65. Zhang, H., Sun, Z., Liu, Z., Xu, C., Wang, L.: Dart: a geographic information system on hadoop. In: Proceedings of the IEEE 8th International Conference on Cloud Computing, pp. 90–97. IEEE (2015)

66. Zhang, J., Mamoulis, N., Papadias, D., Tao, Y.: All-nearest-neighbors queries in spatial databases. In: Proceedings of the 16th International Conference on Scientific and Statistical Database Management, pp. 297–306. IEEE Computer Society, Washington, DC (2004)

67. Zhang, S., Han, J., Liu, Z., Wang, K., Feng, S.: Spatial queries evaluation with MapReduce. In: Proceedings of the 8th International Conference on Grid and Cooperative Computing, pp. 287–292. IEEE Computer Society, Washington, DC (2009)

68. Zhong, Y., Han, J., Zhang, T., Li, Z., Fang, J., Chen, G.: Towards parallel spatial query processing for big spatial data. In: Proceedings of the 2012 IEEE 26th International Parallel and Distributed Processing Symposium Workshops & PhD Forum, pp. 2085–2094. IEEE Computer Society, Washington, DC (2012)

A Cloud-Based Time-Dependent Routing Service

Kalliopi Giannakopoulou[1,2], Spyros Kontogiannis[2,3],
Georgia Papastavrou[2,3], and Christos Zaroliagis[1,2(✉)]

[1] Department of Computer Engineering and Informatics,
University of Patras, 26504 Patras, Greece
{gianakok,zaro}@ceid.upatras.gr
[2] Computer Technology Institute and Press "Diophantus", 26504 Patras, Greece
gioulycs@gmail.com
[3] Department of Computer Science and Engineering,
University of Ioannina, 45110 Ioannina, Greece
kontog@cse.uoi.gr

Abstract. We present a cloud-based time-dependent routing service, which is a core component in providing efficient personalized renewable mobility services in smart cities. We describe the architecture of the time-dependent routing engine, consisting of a core routing module along with the so-called urban-traffic knowledge base, which creates, maintains and stores historic traffic data, as well as live traffic updates such as road blockages or unforeseen congestion. We also provide the crucial algorithmic details for providing the sought efficient time-dependent routing service. Our cloud-based time-dependent routing service exhibits an excellent practical behavior on a diversity (w.r.t. to scale *and* type) of real-world road networks.

1 Introduction

The development of efficient route planning services for traveling in smart cities is a highly sought-after commodity nowadays. Such services are delivered to the smartphones of travelers, who pose route planning queries and receive answers in these devices.

Nevertheless, real-world road networks are typically of very large scale, and demonstrate a time-varying behavior. For instance, the traversal-times of the road segments in the network depend strongly on the actual times of the traversal. This in turn makes it hard, if not impossible due to the practically prohibitive storage and computing requirements, for advanced time-dependent (TD) route planning services to run on isolated devices that have limited computational capabilities, such as our portable navigation devices (PNDs) and smartphones. Additionally, even for a typical route planning server, classical route planning algorithms, such as Dijkstra's algorithm, are not an option. Such a server would have to respond to several dozens, or even hundreds of queries per minute, and Dijkstra's algorithm would require a few seconds per query

© Springer International Publishing AG 2017
T. Sellis and K. Oikonomou (Eds.): ALGOCLOUD 2016, LNCS 10230, pp. 41–64, 2017.
DOI: 10.1007/978-3-319-57045-7_4

for large-scale, time-dependent instances of road networks, thus making such an approach highly impractical. For this reason, extremely efficient heuristic approaches (*speedup* techniques, see e.g., [2] and references therein) and approximation algorithms with provable guarantees (*distance oracles*, see e.g., [1,9–15] and references therein) have been designed, analysed and experimentally tested during the last years, also for time-dependent instances [5–7].

Another axis of complexity, for providing route plans in time-dependent road networks, is the fact that the historic traffic data and, most importantly, the traffic metadata which are created by the routing service, are typically extremely demanding in terms of computational capabilities. Moreover, rather than being created and stored only once, they also have to be periodically updated (say, on a weekly basis), according to the aggregated traffic information of actual speed samples provided by the connected travelers to the service. This is an extremely demanding task for a single server. On the other hand, this maintenance task is extremely parallelizable and of varying computational demands, based on the required changes for the updates. Therefore, the *elasticity of a cloud architecture* would allow for the adaptation of the reserved computing resources to the actual demands for preprocessing traffic-related data and metadata.

One more complication is posed by the fact that the characteristics of the real-world road networks, apart from demonstrating a predetermined time-dependent behavior, also have to cope with unpredicted incidents (e.g., temporal blockages of road segments due to construction works, accidents, etc.), which are typically reported by several sources of information, such as the municipality, the police, or even the travelers themselves. This *live-traffic* information has to be interleaved with the historic (time-dependent) traffic information, in order for the time-dependent routing (TDR) service to provide live-traffic aware routes to the travelers.

All these crucial challenges necessitate the adoption of more sophisticated TDR architectures, which are able to both digest very large amounts of historic information, and also continuously interact with live-traffic sources of information in real time. The heart of such a sophisticated TDR service would have to lie on a cloud architecture, which would be able to guarantee data persistence, interoperability with other sources of traffic-related information, real-time elasticity of computing resources, and also transparent accessibility of the TDR service by the travelers. In such an environment, the queries are sent to a routing engine residing at the cloud infrastructure, which in turn sends back the answers taking into account the updated historic and live-traffic information which is at its disposal.

In this paper, we make two contributions: (i) we describe the architecture of a cloud-based TDR engine that consists of a core module, the so-called *urban-traffic knowledge base* (UTKB), whose role is to create and periodically maintain historic traffic data and metadata, and also to digest in real-time live-traffic update data that are spontaneously provided by diverse sources of information; and (ii) we provide the algorithmic details for providing the sought-after efficient TDR service that exploits all this periodically processed traffic-related information along with the live-traffic updates, in order to respond in real-time to arbitrary route planning queries.

The specific cloud-based architecture of our TDR service constitutes part of a broader cloud-based platform, developed in the frame of [8], whose aim is to provide a live *community of travelers*, equipped with an arsenal of interoperable personalized renewable mobility services, for large-scale urban road networks. Extensive experimentation with real-world data sets (road networks of Berlin and Germany) demonstrated an excellent performance of the core TD algorithmic routing engine [5]. The specific cloud-based TDR service has been also tested in a pilot phase (in the frame of [8]) in the city of Vitoria-Gasteiz, demonstrating very efficient practical behavior.

The rest of the paper is organized as follows. Section 2 presents the formal problem setting along with the necessary definitions and notation. Section 3 presents the architecture of the TDR service that involves the details of the TDR algorithmic engine, the UTKB, as well as the digestion of unforeseen live-traffic (e.g., emergency) reports and traffic prediction alerts. Section 5 presents the results of the application of our TDR service on a real-world environment. We conclude in Sect. 6.

2   Preliminaries

In this section, we provide the necessary definitions and notation that will be used throughout the paper adopted from [6,7]. For any integer $k \geq 1$, let $[k] = \{1, 2, \ldots, k\}$. We consider the classical modeling of a road network as a directed graph $G = (V, A)$, with $|V| = n$ nodes or vertices, and $|A| = m \in \mathcal{O}(n)$ arcs (as is the typical case of such networks). Nodes represent road crossings and an arc $a = (u, v)$ between two nodes u and v represents a road segment between two road crossings (without any other crossing intervening between them).

Every arc $a \in A$ is accompanied with a continuous, periodic, pwl *arc-traversal time* function defined as follows: $\forall k \in \mathbb{N}, \forall t \in [0, T)$, $D[a](kT + t) = d[a](t)$, where $d[a] : [0, T) \to [1, M_a]$ such that $\lim_{t \uparrow T} d[a](t) = d[a](0)$, for some fixed integer M_a denoting the maximum possible travel time ever seen at arc a. Let also $M = \max_{a \in A} M_a$ denote the maximum arc-traversal time ever seen in the entire network. The minimum arc-traversal time value ever seen in the entire network is also normalized to 1. Since every $D[a]$ is periodic, continuous and pwl function, it can be represented succinctly by a number K_a of breakpoints defining $d[a]$. Let $K = \sum_{a \in A} K_a$ denote the number of breakpoints to represent all the arc-traversal time functions in G, $K_{\max} = \max_{a \in A} K_a$, and let K^* be the number of *concavity-spoiling* breakpoints, i.e., the ones in which the arc-delay slopes increase. Clearly, $K^* \leq K$, and $K^* = 0$ for *concave* arc-traversal time functions.

The *arc-arrival-time* function of $a \in A$ is defined as $Arr[a](t) = t + D[a](t)$, $\forall t \in [0, \infty)$. The *path-arrival-time* function of a path $p = \langle a_1, \ldots, a_k \rangle$ in G (represented as a sequence of arcs) is the composition of the arc-arrival-time functions for the constituent arcs:

$$Arr[p](t) = Arr[a_k](Arr[a_{k-1}](\cdots(Arr[a_1](t))\cdots)).$$

The *path-travel-time* function is then $D[p](t) = Arr[p](t) - t$. For any pair of vertices $(o, d) \in V \times V$, let $\mathcal{P}_{o,d}$ be the set of *od*-paths in G.

The *earliest-arrival-time* and *shortest-travel-time* functions are defined as follows: $\forall t_o \geq 0$,

$$Arr[o,d](t_o) = \min_{p \in \mathcal{P}_{o,d}} \{Arr[p](t_o)\}$$

$$D[o,d](t_o) = \min_{p \in \mathcal{P}_{o,d}} \{D[p](t_o)\} = Arr[o,d](t_o) - t_o.$$

The set $SP[o,d](t_o) = \{p \in P_{o,d} : Arr[p](t_o) = Arr[o,d](t_o)\}$ is the set of shortest-travel-time paths for the *query* (o, d, t_o).

A $(1 + \varepsilon)$-*upper-approximation* $\overline{\Delta}[o,d]$ and a $(1 + \varepsilon)$-*lower-approximation* $\underline{\Delta}[o,d]$ of $D[o,d]$, are continuous, pwl, periodic functions, with a (hopefully small) number of breakpoints in $[0, T)$, such that the following inequalities hold: $\forall t_o \geq 0,\ \frac{D[o,d](t_o)}{1+\varepsilon} \leq \underline{\Delta}[o,d](t_o) \leq D[o,d](t_o) \leq \overline{\Delta}[o,d](t_o) \leq (1+\varepsilon) \cdot D[o,d](t_o)$.

3   Architecture of the TDR Service

The *time-dependent routing* (TDR) service aims at supporting the TD route-planning functionality for vehicles (be it conventional cars, or electric vehicles), where the typical optimization criterion is the minimization of the total travel-time. In particular, the TDR service is responsible for executing several types of queries made by users. Based on the user requirements or/and preferences, it computes one or several routes which satisfy at least one optimization criterion (such as distance, travel time, fuel consumption, eco-friendliness) and use at least one transportation mode (bus, train, EV, bicycle), at specific departure times or time windows.

An overview of the overall architecture of the TDR service, along with its interaction with other services of the broader cloud system of [8], is depicted in Fig. 1. The TDR service consists of data and metadata structures, mechanisms for creating and maintaining these structures, as well as algorithms that answer user route planning requests in real-time by exploiting the stored data and metadata.

In particular (cf. Fig. 1), the *raw-traffic data* (RTD) is a collection of sequences of breakpoints, one per arc in the network. Each breakpoint is a pair of departure-time (from the tail) and traversal-time (up to the head). The *urban-traffic knowledge base* (UTKB) consists of mechanisms for creating and maintaining the RTD structure. Apart from that, in order for the travelers' PNDs and smartphones to be able to provide elementary route plans even when there is no connection to the cloud, the UTKB aggregates the RTD structure into (static) traffic snapshots (SNAP), which are to be stored in the travelers' devices, as a means of contingency plan in case of loss of connectivity. In support of the sophisticated route planning algorithms (cf. Sect. 4.2), relevant traffic

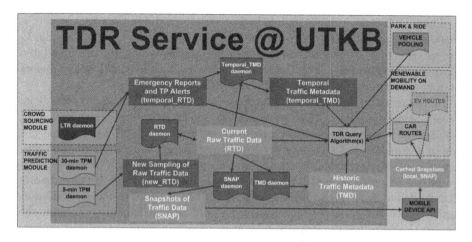

Fig. 1. Overall architecture of the Time-Dependent Routing (TDR) service.

metadata (TMD) is created and periodically updated. This structure concerns the succinct representation of the (approximate) min-travel-time functions from selected vertices (called *landmarks*) to reachable destinations from them. The UTKB also provides the appropriate procedures to digest live-traffic reporting, i.e., for creating and maintaining temporal RTD and temporal SNAP structures. These are succinct representations (i.e., sequences of breakpoints) of approximate min-travel-time functions from landmarks, which are only effective for a given time-window (depending on the duration of the reported incident). Last but not least, the TDR service also hosts the core TDR query algorithms (its most vital part) that exploit all the aforementioned traffic-related data and metadata.

As mentioned above, all these TDR functionalities, be it data-maintenance procedures or route-planning query algorithms, constitute the urban traffic knowledge base (UTKB) which resides on one or more computing resources of the cloud, depending on the actual computational demands of these functionalities. In the rest of this section, we provide the architectural details of UTKB.

3.1   Urban Traffic Knowledge Base

All the time-dependent urban traffic information and live-traffic monitoring information is organized and maintained in a periodically updated and dynamically-evolving urban-traffic knowledge base (UTKB), so as to support responses to route-planning queries in real time. The UTKB is responsible for the creation and maintenance of traffic-related metadata, to be exploited by route planning and mobility-on-demand services supported by the overall cloud platform. Its main purpose is to handle the periodically changing urban traffic information, by dynamically updating the preprocessed traffic data kept in the system, when needed.

The UTKB receives input mainly by two system modules, the *Crowd-sourcing Module* (CM) and the *Traffic Prediction Module* (TPM), which are used as black-box services and are responsible for creating and assessing the traffic-related information received either by the road network itself, or by the travelers, before sending appropriate update signals to the UTKB.

More precisely, the incoming data may concern *periodic traffic reports* (e.g., traversal-time samples of road segments, every 5 min), *emergency reports* (e.g., spontaneous reports of accidents, predictions of unforeseen evolution of congestion in particular road segments, etc.), information on changes of external parameters (e.g., weather conditions), updates on public-transport's mobility plans, or updates on publicly used EVs' availability information. Each of these reports actually demands for an appropriate update of the involved traffic-related information. Both the range of affection of a newly reported incident (within the network) and the temporal traffic-related metadata in response to this particular report, has to be determined by the UTKB itself. As a result, the historic traffic data and metadata kept within the UTKB is interleaved with the temporal traffic metadata created per reported incident, but only for those routes which are indeed affected by it, so that live-traffic aware responses to arbitrary route-planning queries are provided by the query algorithms in real time.

As previously mentioned, the UTKB also creates and periodically maintains snapshots of the current traffic status (that is, average arc-traversal time *values* rather than time-dependent arc-traversal time *functions*, e.g., for rush-hours or free-flow route plans, depending on the traveler's departure time), to be at the disposal of the travelers for downloading them to their personal devices, so as to assure a minimum level of the routing service even without connection to the cloud system.

The main purpose of the UTKB is to gather and digest all the (spontaneous, or periodic) observations of the live-traffic situation, and dynamically update its contents. The input data and the corresponding update actions applied to them can be divided into seven basic categories: (i) user generated periodic speed reports, mainly concerning private cars and EVs; (ii) information on public-transport data (that includes static information, such as timetables, but also planned events, e.g. a subway station must be closed due to maintenance, as well as live traffic data, where in particular, delays are important); (iii) energy-consumption information (e.g., current state-of-charge) concerning EVs; (iv) spontaneously provided emergency reports, which include unpredictable traffic disruptions (i.e., currently unavailable road segments, changes on weather conditions, etc.), reported either by a (totally reliable) public authority, or by (possibly unreliable) travelers whose credibility is based on a wisdom-of-the-crowd approach (the more travelers reporting an incident, the more likely it is that it is actually true); (v) short-term traffic prediction alerts, as reported by the TPM, acting as periodic travel-time samples (say, every 5 min) of the entire network; (vi) long-term traffic prediction alerts, as reported by the TPM (say, every 30 min), in order to automatically detect unforeseen evolution of the traffic pattern and provide appropriate signals (analogous to the emergency reports)

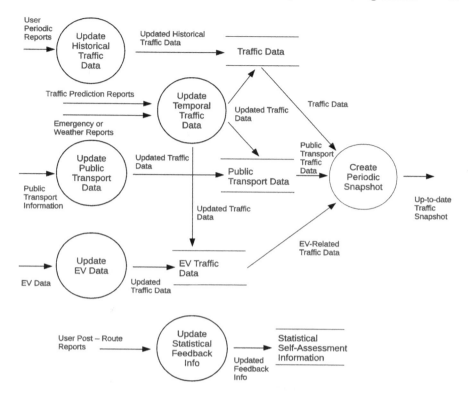

Fig. 2. Detailed DFD for the UTKB architecture.

whenever the predicted behavior of a road segment deviates significantly from the behavior described in the historic raw data; and (vii) users' post-route spontaneous reports, containing route-related information, or a kind of like/dislike feedback on the routes recommended to them, as a means of self-assessment of the TDR service.

In each case, an appropriate update on the historical or the temporal traffic data and metadata kept in the system must be performed by the UTKB. A corresponding bubble, describing each update process, is added in a detailed data flow diagram (DFD) of the UTKB, shown in Fig. 2. All the aforementioned types of information are carefully collected and stored in the UTKB, therefore a data storage is displayed for each type of data. Additionally, a unique process should take over the responsibility of periodically creating a snapshot of the current traffic status, which will be available for downloading to the users' portable devices, upon their own request.

A second-level refinement of UTKB's architecture is accomplished by mapping individual bubbles or groups of bubbles on the same side of a boundary of the DFD into appropriate modules within the UTKB's architecture. Four main submodules of the UTKB module are considered: (i) Historical Traffic Data Update; (ii) Temporal Traffic Data Update; (iii) Statistical Feedback Information Update;

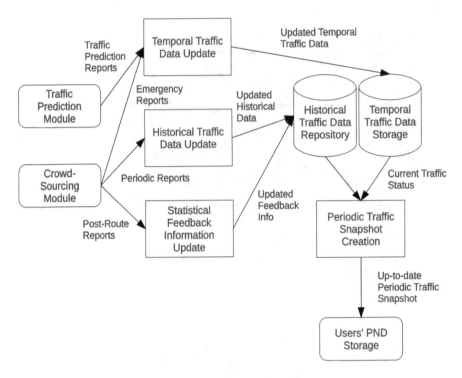

Fig. 3. The high-level architecture diagram for UTKB Module.

and (iv) Periodic Traffic Snapshot Creation. The data that the UTKB module keeps and handles, either historical or temporal, may concern all types of transport modes, be it private cars, EVs or public-transport modes. Based on the resulting sub-modules, the corresponding hierarchical schema turns out to be too simple. The result is depicted in Fig. 3, which shows the high-level architectural diagram of the UTKB.

4 Functionality of the Cloud-Based TDR Service

In this section, we describe the functionality of the TDR service for computing routes in road networks, concerning mainly private cars.

4.1 Creation and Maintenance of Traffic Metadata

We start with the algorithmic technique used for the actual creation of the traffic metadata (TMD) concerning private cars, as well as the methodology adopted for the efficient maintenance of all the traffic-related information kept in the UTKB, so that it can be continuously available and exploited by the TDR service residing at the UTKB server. The theoretical analysis (correctness and complexity bounds) of the algorithmic technique is provided in [6].

The TDR service is able to respond in real time to arbitrary shortest-path queries, by computing an optimal origin-destination path providing the earliest-arrival time at a destination, when departing at a given time from the origin. In order for the route-plans given by TDR to be fast and accurate, the service exploits the appropriately selected shortest-path information, precomputed off-line. In particular, a carefully selected subset of vertices (*landmarks*) is equipped with succinct representations of shortest-travel time *functions* to all other vertices in the network. Apart from creating these traffic metadata, the UTKB has also to update them periodically, according to the periodical adaptations of the historic traffic data kept in it. Both the creation and the periodic updating of all this landmark-related traffic metadata, are extremely time-consuming tasks whose exact computational needs are indeed unclear.

The role of the cloud, which allocates the appropriate amount of processing power depending on the work to be done, is actually crucial at this preprocessing phase which has to be completed within a few hours and is repeated frequently (e.g., once per week). Since computing and storing the exact shortest-travel time functions turns out to be hard (in particular, super-polynomial) [3], we compute $(1 + \varepsilon)$-approximations of these functions, called *travel-time summaries*, from the selected landmarks towards all other vertices. The main challenges of this major task are: (i) to ensure that the preprocessing time and space complexity is actually manageable (e.g., subquadratic in the size of the network); (ii) to allow for route-planning query algorithms which provide fast responses, both in theory and in practice; and (iii) to obtain provably good approximation guarantees.

Approximating Travel-Time Summaries. Our main building block for the preprocessing is an approximation technique for the computation of all landmark-to-vertex approximate travel-time summaries. We briefly describe here a novel method, the *trapezoidal technique* (TRAP), which is a one-to-many method that concurrently computes travel-time summaries from a given land-mark to many (or even all) destinations which are reachable from it.

For a given set of landmark-nodes L (whose choice will be determined later), our goal is to construct all the $(1 + \varepsilon)$-upper-approximation shortest travel-time functions (travel-time summaries) from each landmark towards all reachable destinations, for a time period of a day, i.e., in the interval $[0, T = 86400\,\mathrm{s})$.

Instead of computing the exact minimum-travel-time (which are continuous, pwl and periodic) functions $D[\ell, v]$ from each $\ell \in L$ towards each reachable $v \in V$, we seek for their $(1 + \varepsilon)$-upper-approximations $\overline{\Delta}[\ell, v]$ (the travel-time summaries). Recall from Sect. 2 that $\overline{\Delta}[\ell, v]$ and $\underline{\Delta}[\ell, v]$ denote $(1 + \varepsilon)$ upper- and lower-approximations of $D[\ell, v]$ which are also continuous, pwl, periodic functions, hopefully with a small (in particular, independent of the size of the network) number of breakpoints in $[0,\mathrm{T})$, such that the following inequalities hold: $\forall t_\ell \geq 0$,

$$\frac{D[\ell, v](t_\ell)}{(1 + \varepsilon)} \leq \underline{\Delta}[\ell, v]](t_\ell) \leq D[\ell, v](t_\ell) \leq \overline{\Delta}[\ell, v](t_\ell) \leq (1 + \varepsilon)D[\ell, v](t_\ell)$$

The *trapezoidal method* (TRAP) is a one-to-all approximation algorithm for computing concurrently all functions $\overline{\Delta}[\ell, v]$ for a given landmark ℓ and all reachable destinations $v \in V$. The theoretical analysis of TRAP can be found in [6]. TRAP splits the entire period $[0, T)$ in small (length-τ) subintervals, and within each of them, say $[t_s, t_f = t_s + \tau) \in [0, T)$, it provides the appropriate projection of $\overline{\Delta}[\ell, v]$, by essentially exploiting the fact that $\tau > 0$ is indeed small, along with the following assumption on the travel-time slopes of all minimum-travel-time functions in the network:

Assumption 1 (Bounded Travel-Time Slopes). All min-travel-time slopes are bounded in a given interval $[-\Lambda_{\min}, \Lambda_{\max}]$, for $\Lambda_{\min} \in [0, 1)$ and $\Lambda_{\max} \geq 0$.

The validity of this assumption has been experimentally verified in real-world data sets that we have at our disposal [5].

Within each subinterval $t_s, t_f = t_s + \tau)$, TRAP provides a crude approximation of the unknown function $D[\ell, v]$, concerning the minimum slope $-\Lambda_{\min}$ and maximum slope $\Lambda_{\max}$ of the actual shortest-travel-time functions in the instance. In particular, for every subinterval $[t_s, t_f)$, TRAP works as follows. For the two boundary departure times t_s and t_f, we sample concurrently (by making two calls to the time-dependent variant of Dijkstra's algorithm) the travel-time values for each destination $v \in V$. We then consider the semi-lines with slope $\Lambda_{\max}$ from t_s and $-\Lambda_{\min}$ from t_f. The upper-approximating function $\overline{\Delta}[\ell, v]$ that we consider within $[t_s, t_f)$ is the lower-envelop of these two semi-lines. Analogously, the lower-approximating function $\underline{\Delta}[\ell, v]$ is the upper-envelop of the semi-lines that pass through t_s with slope $-\Lambda_{\min}$, and from t_f with slope $\Lambda_{\max}$. In particular, TRAP considers the following two (upper- and lower-) approximating functions of $D[\ell, v]$ for every possible departure time $t \in [t_s, t_f)$:

$$\overline{\Delta}[\ell, v](t) = \min \left\{ D[\ell, v](t_s) - \Lambda_{\max} t_s + \Lambda_{\max} t, D[\ell, v](t_f) + \Lambda_{\min} t_f - \Lambda_{\min} t \right\}$$

$$\underline{\Delta}[\ell, v](t) = \max \left\{ D[\ell, v](t_f) - \Lambda_{\max} t_f + \Lambda_{\max} t, D[\ell, v](t_s) + \Lambda_{\min} t_s - \Lambda_{\min} t \right\}$$

Considering $\overline{\Delta}[\ell, v]$ as the required travel-time summary for departure-times in $[t_s, t_f)$, the algorithm has to decide whether this is actually a $(1 + \varepsilon)$-upper-approximation of $D[\ell, v]$. For this reason, we must compute the maximum additive error $MAE(t_s, t_f)$, which is the maximum delay-axis distance of the two functions $\overline{\Delta}[\ell, v]$ and $\underline{\Delta}[\ell, v]$ within $[t_s, t_f)$. This is done as follows. Let $(\underline{t}_m, \underline{D}_m)$ be the intersection point of the two legs involved in the definition of $\underline{\Delta}[\ell, v]$. Similarly, let $(\overline{t}_m, \overline{D}_m)$ be the intersection point of the two legs involved in the definition of $\overline{\Delta}[\ell, v]$. The (worst-case) *maximum additive error* $MAE(t_s, t_f)$ guaranteed for $\overline{\Delta}[\ell, v]$ within $[t_s, t_f)$ is:

$$
\begin{aligned}
MAE(t_s, t_f) &:= \max_{t \in [t_s, t_f)} \left\{ \overline{\Delta}[\ell, v](t) - \underline{\Delta}[\ell, v](t) \right\} \\
&= \overline{\Delta}[\ell, v](\underline{t}_m) - \underline{\Delta}[\ell, v](\underline{t}_m) \\
&= \overline{\Delta}[\ell, v](\overline{t}_m) - \underline{\Delta}[\ell, v](\overline{t}_m)
\end{aligned}
$$

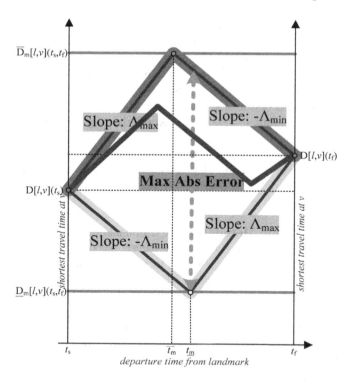

Fig. 4. The upper-approximating function $\overline{\Delta}[\ell, v]$ (the orange, upper pwl line), and lower-approximating function $\underline{\Delta}[\ell, v]$ (the yellow, lower pwl line), of the unknown distance function $D[\ell, v]$ within $[t_s, t_f)$. (Color figure online)

Figure 4 provides a visualization of all the above mentioned quantities, as well as the upper- and lower-approximating functions returned by TRAP within $[t_s, t_f)$.

For each subinterval $[t_s, t_f)$, the algorithm checks whether the constructed upper-approximating function $\overline{\Delta}[\ell, v]$, valid for every possible departure time $t \in [t_s, t_f)$ from the origin ℓ, is actually a $(1 + \varepsilon)$-upper-approximation of the exact shortest-path function in the same interval. If this is the case, $\overline{\Delta}[\ell, v][t_s, t_f)$ is accepted. Otherwise, the length of the sampling interval τ needs to be even smaller. TRAP handles all possible sampling intervals as follows.

Rather than splitting the entire period $[0, T)$ in a flat manner, i.e., into equal-size intervals, we start with a coarse partitioning based on a large length τ and then in each interval and for each destination vertex we check for the provided approximation guarantee by TRAP. All the vertices which are already satisfied by this guarantee with respect to the current interval, become inactive for this and all its subsequent subintervals. If there is at least one active destination vertex, for which the function $\overline{\Delta}[\ell, v]$ constructed in the current interval violates the maximum absolute error, we proceed by splitting in the middle the current subinterval, and repeat the check within the new subintervals created. The algorithm terminates when all reachable destinations become inactive for every subinterval of $[0, T)$, which means that every one of them has a $(1 + \varepsilon)$-upper-approximation function for the entire period.

We keep every constructed $\overline{\Delta}[\ell, v]$ function as a sequence of sampled breakpoints, which are of the form (*departure-time, travel-time, approximate-path-predecessor*). The predecessor of v results from the sampling that the algorithm performs at each subinterval $[t_s, t_f)$. To reduce the required space in memory, we do not store the node-id of each predecessor. Instead, we only need to keep a small integer indicating the index of the corresponding incoming arc of v in the adjacency list. We have further developed a host of heuristic algorithmic improvements to boost performance in practice [5]; for completeness, we describe them in Appendix A.

Maintenance of Traffic Metadata. The traffic metadata produced by TRAP are efficiently maintained in the UTKB server, according to the following methodology. In order to access the data blocks of each landmark in $O(1)$ time, we use a mapping. For retrieving efficiently each approximate travel-time function from a landmark to any destination vertex, we need to store an index. In particular, for each landmark ℓ we maintain a vector of pointers, the size of which equals to the number of destinations. The order of the pointers is in ascending order of node id and each one of them corresponds to a destination v. Thus, the address of the $\overline{\Delta}[\ell, v](t)$ data is provided in $O(1)$ time, while the required space for this indexing is $O(n \cdot |L|)$.

To keep the preprocessing space small, we store the information about both predecessors using the same unit of memory. In the case that the approximate function from a landmark ℓ to a destination v is constant, we expect the approximate predecessor of v to be the vertex ℓ and the predecessor corresponding to the piecewise composition (not performed in this case) to be the destination vertex v itself.

We describe next how the appropriate traffic metadata related to every day of the week are uploaded in the UTKB server, so that they are available to the TDR service. A continuously running TDR daemon is responsible for this task. In particular, at the beginning of each day the corresponding traffic-data is uploaded, while the data related to the previous day is automatically removed.

Finally, the TDR daemon also undertakes the creation of new updated TMD for any day in an off-line fashion (e.g., at the end of each day). In particular, the TPM module (used as a black-box daemon service, also residing at the cloud) provides the TDR daemon with periodic (say, per 5 min) predictions of travel-time value estimations, for all the road segments in the network. All these samples constitute, by the end of the day, a fresh image of the historic data for the entire day, which is then aggregated by the TDR daemon in the current historic data (with an appropriately small weight, so as to avoid over-fitting or oscillation effects), so that the historic raw traffic data residing at the UTKB converge to the actual travel-time functions of all the road segments, per day.

4.2   TDR Query Algorithms

The TDR service is developed to provide significantly fast and accurate min-cost route plan responses to arbitrary shortest-path queries, exploiting (i) a carefully selected landmark-set of vertices; (ii) the historical and temporal traffic-related information kept in the UTKB server; and (iii) an efficient approximate query algorithm, designed to provide the required routes. In this section, we describe how the query algorithms supported by the routing service work, as well as how the exploitation of the historical traffic-data and those provided by the Traffic Prediction Module and the Crowd-sourcing Module is achieved.

The daemon residing at the UTKB server continuously runs and accepts incoming origin/destination/departure-time shortest-path queries (o, d, t_o). For each one of them, the query algorithm provided by the routing service is called and returns either the exact minimum-travel-time value, along with the corresponding o-d path, or an approximate travel-time value via an appropriate landmark-node ℓ and the corresponding approximate o-ℓ-d path.

Three query algorithms have been implemented and extensively tested. We describe them below. Their theoretical analysis (correctness and complexity bounds) is provided in [6,7].

The first one, which is called FCA, grows a Time-Dependent-Dijkstra (TDD) ball from (o, t_o) until either d or the closest landmark l_o is settled. In the former case it returns the minimum travel-time value and the corresponding shortest path. In the latter case, the $(1 + \varepsilon + \psi)$-approximate travel-time value of an o-d path passing by l_o is returned, where ψ is a constant that depends on characteristics related with travel-times, but is independent of the size of the network.

The second algorithm, called $FCA^+(N)$, is a variant of FCA which keeps growing a TDD ball from (o, t_o) until either d or a given number N of landmarks is settled. FCA^+ then returns the exact travel-time value, or the smallest via-landmark approximate travel-time value, among all these settled landmarks. Theoretically, the approximation guarantee of FCA^+ is the same as that of FCA, but its practical performance is actually impressive [5].

The third algorithm, called RQA, improves the approximation guarantee provided by FCA, by exploiting carefully a number r (called the *recursion budget*) of recursive accesses to the preprocessed information, each of which produces (via calls to FCA) additional candidate o-d paths. RQA works as follows. As long as the destination vertex within the explored area around the origin has not yet been discovered, and there is still some remaining recursion budget, it "guesses" (by exhaustively searching for it) the next vertex w_k of the boundary set of touched vertices (i.e., still in the priority queue) along the unknown shortest o-d path. Then, it grows an outgoing TDD ball from every new center $(w_k, t_k = t_o + D[o, w_k](t_o))$, until it reaches the closest landmark ℓ_k to it, at travel-time $R_k = D[w_k, \ell_k](t_k)$. Every new landmark offers an alternative o-d path by a new application of FCA for every boundary center w_k. RQA finally responds with a $(1 + \sigma)$-approximate travel-time to the query, for any constant $\sigma > \varepsilon$.

The response-times as well as the approximation guarantees provided by all three query algorithms have a strong dependence on the selected landmark-set [5]. Our experimental study [5] has shown that FCA has the fastest performance (as expected), and provides quite small approximation guarantees. Both RQA and FCA$^+$ are fast, i.e., they run in time less than 1 msec, but also significantly accurate, since they produce solutions with relative errors less than 1% in the average case. FCA$^+$ is in some cases better than RQA regarding accuracy, while it is almost as fast as RQA. In fact, we can control the trade-off between time and accuracy, by selecting a smaller or larger number of landmarks to be discovered by the algorithm. For those reasons, FCA$^+$ is the default algorithm running in the TDR service.

Live-Traffic Awareness and Route Computation. We describe in detail the way in which the query algorithm manages to exploit the historical traffic-data as well as the temporal data corresponding to live-traffic updates, and finally provides the resulting route as output, after the application of a path reconstruction method, for retrieving the unknown approximate paths.

For any incoming shortest-path query, the algorithm considers the flags associated with all arcs and landmarks in the network, to indicate whether there exists active temporal data for them or not. Any temporal traffic-updates have to be adopted. As described in Sect. 4.3, if there are any road segments (and the relevant raw-traffic data), or landmark-nodes (and the associated traffic metadata maintained by the UTKB) affected by an emergency report or traffic prediction alert, a flag corresponding to those arcs and landmarks in the network is raised, to indicate that for a specific time-window the temporal RTD and temporal TMD structures have to be considered, rather the original historic information kept in the RTD and TMD structures, respectively. The query algorithm absorbs the live-traffic changes, by taking into account these flags on the affected arcs and landmarks.

More precisely, we consider the two basic phases of any query algorithm, i.e., the first step, which is a TD Dijkstra-based search, and the second one, which retrieves the approximate distance from a settled landmark-vertex to the destination, by searching into the appropriate preprocessed distance function. For the first step, we perform the following modification on the relaxation of arcs. Each time that we need to compute the travel-time needed to traverse an arc, given a departure-time from its tail, the algorithm checks the flag corresponding to the arc, so as to search for its travel-time function either in the current or the temporal RTD. We need to take the temporal raw traffic-data into account, in the case that a specific arc was recently affected by a live-traffi update, for a particular time-window around the departure time from its tail. The update-daemon running on the UTKB server is responsible to modify the temporal RTD for all affected arcs and raise the bit flag on them so long as an update has been adopted and is still active. Both the historic and the temporal travel-time functions are stored per arc.

The second phase of the algorithm adopts a similar modification for landmarks. If the destination was not discovered by the first step, the algorithm collects one or more landmarks, each providing a distinct approximate solution towards the destination. If a landmark-node has been affected by a live-traffic update, we have to store the time window of corresponding departure times from it for which the update will still be active, as well as a pointer to the address of the corresponding temporal TMD for this landmark. For each landmark, the algorithm checks whether its update-flag bit is 0 or not, so as to decide if the specific landmark has active temporal traffic-data for a particular time window, and therefore there exists an updated approximate distance function from it, kept at the appropriate memory block. The temporal TMD are considered for a landmark ℓ when its update-flag bit is still 1 upon arrival at ℓ, meaning that there is a pointer to a memory address which is not NULL, and the departure time from ℓ is within an affected time window.

By adopting these simple modifications, we are able to exploit the real-time traffic conditions of the network and provide fast and accurate responses to route-planning requests.

Finally, we describe the *Path Reconstruction* method followed for the generation of the computed o-d path, as a sequence of arcs. In the case that the shortest path returned by the query algorithm is exact, i.e., the destination was discovered during the TDD-search (which is actually quite possible to occur), the path is constructed by simply following the predecessors from the destination to the origin, which the TDD ball provided. However, if the algorithm decides that the destination is to be reached via an appropriate landmark ℓ, we need a method to retrieve the unknown sequence of predecessors from the destination up to the landmark, corresponding to the approximate ℓ-d path. For this purpose, we exploit the information kept in all TMD. For every possible destination d and for any departure time from ℓ, the travel-time function contains the immediate predecessor of d, valid for a specific time interval, where TRAP performed its sampling. Based on this information, along with some heuristic ideas, the path reconstruction works as follows.

Let v denote every node that belongs on the path we want to construct. We start with $v = d$. We then obtain the immediate (approximate) predecessor, $approxPred(v)$, given by the approximate travel-time function $D[\ell, d](t)$, when departing from ℓ at time $t_\ell = Arr[o, \ell](t_o)$, which denotes the arrival time set by the Dijkstra-ball. We then mark node v as visited, we set $v = approxPred(v)$ and repeat. The procedure terminates when we reach the landmark ℓ, i.e., $v = \ell$, or at least some already settled vertex by the first time-dependent Dijkstra (TDD) ball grown from (o, t_o). The retrieval of all predecessors is done by searching either the historical or the temporal TMD, depending on the flag that the algorithm previously considered for ℓ.

In practice, we observed the following phenomena which we tackled accordingly. Firstly, as we reversely approach the landmark ℓ, the sequence of nodes v at some point enters the area explored by the query algorithm. We decided to collect all those explored nodes v and compute the total travel-time $D[v, d](t_v)$, exploiting the reverse arc-traversal time functions on all arcs connecting all nodes v up to that point. When the main loop of our method terminates, we check which explored node on the path we constructed (including the landmark ℓ) gives the minimum $D[o, v](t_o) + D[v, d](t_v)$. This means that there can be some cases where we construct the approximate path via an appropriate explored node v of the TDD ball and not the proposed landmark ℓ, which mainly acts as an "attractor" during the path reconstruction phase, rather than an actual intermediate node of a candidate o-d path.

Next, we observed that a predecessor given by the travel-time summaries stored in UTKB can in fact be already visited, which means that a cycle is possibly created. This can be expected since we search different approximate functions $\overline{\Delta}[\ell, v](t)$ for each vertex v, departing from ℓ. The TRAP method samples the exact travel-time-function in different subintervals for each destination. We choose to face this case as follows. The path reconstruction method returns to its initial step, where $v = d$. Instead of departing from landmark ℓ at the exact departure time t_ℓ, we seek for the closest departure time to t_ℓ contained as a breakpoint in *all* approximate distance functions of the predecessors involved to an approximate path up to ℓ. This safely means that this departure time is a sampling time for all those destinations and thus, they all belong to the very same shortest-path tree, created by the TRAP method during the preprocessing. To avoid the cycle (which in practice is a rare case), we consider the sequence of vertices created, considering the above departure-time from ℓ, which is usually very close to the actual.

After the sequence of predecessors has been constructed, the method simply walks on the arcs connecting them and the ℓ-v path-travel-time value is provided by the (actual) path-travel-time function, when departing from ℓ at its actual arrival-time, set by the algorithm. The path-travel-time function can be provided by historic or temporal raw traffic-data, depending on the flags kept on each arc. The resulting value is at most the approximate one. In practice, we obtain a much better travel-time value. The last step is to connect the constructed ℓ-v path with the exact o-ℓ path, given by the query algorithm, leading to a total o-d route-response, which is usually very close to the exact one.

4.3   Adaptation to Emergency Reports and Traffic Prediction Alerts

The TDR service is responsible for computing shortest paths with respect to the current status of the network. In such a service that responds to several queries in real-time, various disruptions may occur "on the fly" (e.g., unforeseen congestion, or even blockage of a road segment). In such a case, the new traffic conditions have to be absorbed in real time and they have to be taken into account by the TDR.

The update of the arc-traversal time functions and the landmark travel-time summaries which are used by the query algorithms is assigned to an online update-daemon worker within TDR. The daemon performs three tasks, whose main functionalities are detailed in the following.

Periodic Inspection for Emergency Alerts (EMAs). The arc-traversal time changes in the network are supplied by reports. These are produced by the

TPM and they are stored in a file called `alerts.csv`, within UTKB. The daemon periodically reads the file (every 15 min), in intervals in which the TPM-daemon does not write. For preventing the infrequent case of reading the file when is still being written by the TPM-daemon, its last modification time in the file system is probed. If it is different before and after the reading phase, then the reading phase is repeated after 1 min. If `alerts.csv` is not empty, then the update-daemon loads in TDR the affected arcs which have to be updated.

Network Data Update. The TDR must work continuously in order to answer to any user or service query. On this requirement, the update steps have to be performed independently without interrupting the TDR. Under normal conditions, the arcs which need to be updated are few. Therefore the chance for an update not be absorbed in the shortest path computation is small. Also, in the worst case, since the update can be completed in a few milliseconds, running shortly a new query will eventually output the updated shortest paths.

During the reading phase of the disrupted arcs from `alerts.csv`, the daemon inserts them in a queue. Then, it runs the update process for each such arc $a = (u, v)$. Let $T = (t_s, t_e]$ be the affection interval associated with an affected (closest) landmark ℓ and related to the disruption occurred at arc a. During that interval, the temporal data from ℓ should be taken into account. Let the new traversal-time value along arc a be Δ at the time interval $T = (t_s, t_e]$, and the original arc-traversal time function be

$$
travelTime_a(t) = \begin{cases}
t \cdot \text{slope}_1 + \text{offset}_1, & [t_0, t_1) \\
t \cdot \text{slope}_2 + \text{offset}_2, & [t_1, t_2) \\
\quad \cdots & \cdots \\
t \cdot \text{slope}_k + \text{offset}_k, & [t_{k-1}, t_k) \\
\quad \cdots & \cdots
\end{cases}
$$

The update steps are as follows.

STEP 1: Initially, the daemon inserts the affection expiration time t_e of the new travel time on a in a priority queue PQ. This is because the emergency alerts may concern short-term changes. Consequently, after the expiration of the time-window of affection, the original travel time function of a will be restored.

STEP 2: The daemon generates the updated travel time function of arc a. Initially, it detects the affected legs of $travelTime(t)$ which have to be updated, within the time interval $[t_s, t_e]$. For example, if the new travel time Δ is occurred in $[t_{k-1}, t_k)$, then the candidate legs to be modified are ($\text{slope}_i, \text{offset}_i$), $i = k - 1, k, k + 1$. In such a case, the linear interpolation is applied on the new travel time values throughout the interval $[t_s, t_e]$. The updated travel time function is stored in a different memory address.

In order TDR to be informed about the new travel time function of a, an update-flag bit is associated to a and it is set to 1 only if the creation of the function is finished. Consequently, during a shortest path computation the update-flag bit of a indicates that its travel time function has changed. In addition, if a

belongs to a shortest path, then a pointer to the address of the updated travel time function is accessed by the routing service.

STEP 3: The daemon re-computes the travel-time summaries for a subset of landmarks L in the vicinity of the disrupted arc a. In particular, it runs a backward TD-Dijkstra from u (tail-node of a) under the free-flow metric, with travel time radius of $t_e - t_{cur}$, where t_{cur} is the current time (based on the corresponding network's UTC). The travel-time radius is used to trace only the nearest landmarks that may actually be affected by the disruption, leaving unaffected all the "faraway" landmarks. This means that the update has to be performed only for the involved drivers who are close to the area of disruption. For each affected landmark ℓ, we consider a disruption-times window $T_\ell = [d_s, d_e]$, containing the latest departure times from ℓ for arriving at the tail u at any time in the interval $[t_s, t_e)$ in which the disruption occurs. The T_ℓ windows, for all landmarks $\ell \in L$, are computed by running two backward TD-Dijkstra queries from u under the time-dependent-flow metric, the first with arrival-time equal to t_s and the second with arrival-time equal to t_e. We then compute the temporal travel-time summaries for each affected landmark ℓ at its disruption-times window T_ℓ. The produced travel-time summaries are stored in a different memory address.

In order TDR to be informed about the new travel-time summaries of landmark ℓ, an update-flag bit is associated to ℓ and it is set to 1 only if the creation of the travel time summaries is finished. Consequently, during a shortest path computation, if ℓ is required at the specific disruption-time window T_ℓ, the update-flag bit indicates that the travel time summaries of ℓ have changed on T_ℓ. In this case, a pointer with the address of the updated travel-time summaries is accessed by the routing service.

Expiration Monitoring. The daemon wakes up from the idle state when the affection expiration time t_e of a disrupted arc is reached, based on the network's UTC. In such a case, it extracts the arc a having the earliest t_e from the priority queue PQ. Then it sets the update-flag bit to 0 on arc a and any corresponding landmark $\ell \in L$ which were updated due to a. Consequently, during a shortest path computation, when it is required, the routing service will use the original travel time function of arc a. Similarly, if a landmark $\ell \in L$ is used on a shortest path computation, then the routing service will use the original travel time summaries of ℓ. In the end, the daemon removes the outdated temporal travel time functions and summaries.

5   Practical Performance of the TDR Service

The TDR cloud-service constitutes part of a broader cloud-based platform, developed in the frame of [8], that delivers personalized services for renewable mobility within cities. In this section, we report on the practical performance of TDR. In particular, (i) we report on an extensive experimental study carried out with real-world data sets from the road networks of Berlin and Germany [5]; (ii) we report on a pilot study of the TDR service carried out in real-world conditions in the city of Vitoria-Gasteiz (in the frame of [8]).

5.1 Experimental Study on Real-World Road Networks

In this section, we succinctly report the outcome of our experiments on the instances of Berlin and Germany (details can be found in [5]). The instance of Berlin consists of $473, 253$ nodes and $1, 126, 468$ arcs, while the instance of Germany consists of $4, 692, 091$ nodes and $11, 183, 060$ arcs.

We measured the performance of the basic query algorithms FCA, FCA$^+$ and RQA, with respect to absolute running times and Dijkstra-rank[1] values, respectively, for several types of landmark set selections (for instance, variants of a random selection, and/or partition-based selection of landmarks from the boundary vertices of the partition), and various sizes.

In particular, we considered a uniformly at random selection of landmarks among all vertices, denoted by R, and a random selection of landmarks among the boundary vertices resulted by the graph-partitioning tool KaHIP [4], denoted by K. Two variations of S and K stood clearly above others: the variation SR of R, where each newly chosen random landmark excludes its closest 300 vertices (under the free-flow metric) from being landmarks in the future; and the (similar in nature with SR) variation SK of K.

As for the query algorithms, we used recursion budget 1 for RQA and we let FCA$^+$ settle the 6 closest landmarks, which is roughly the average number of settled landmarks by RQA as well.

For Berlin, our fastest query algorithm, FCA, combined with the SR-landmark set, achieved average response times of $83 \, \mu s$, with relative error of 0.781%, absolute runtime speedup more than 1146, and Dijkstra-rank speedup more than 1570, compared to TDD. If the relative error is of importance, then one should choose FCA$^+$(6) along with the SK-landmark set, which achieves $0.616 \, ms$ query time, and relative error of 0.226%. In case that space is a main concern, we observed the full scalability in the trade-offs between space and query-responses. For instance, by consuming space $3.2 \, GB$, we can achieve query-response time $0.73 \, ms$ and relative error 2.198% for the Berlin instance.

In Germany, our findings are analogous. Exploiting 6 computational threads, the average preprocessing time is less than $90 \, s$ and the average space is up to $25.7 \, Mbytes$, per landmark. The best speedup against TDD is achieved by SK-landmarks, and is more than 1531 in Dijkstra-rank, and more than 902 in absolute query-time, with worst-case error at most 1.534%.

The aforementioned results suggest that our TDR service (and in particular its TDR engine) is suitable for practical application. The has indeed happened and described in the next section.

---

[1] The Dijkstra-rank of a vertex v is the number of settled vertices by (plain or TD) Dijkstra's algorithm until v is settled.

Fig. 5. Earliest-arrival-time query before and after an emergency report.

5.2   Pilot Execution in a Smart City

During 2016, the TDR service has been piloted for more than three months in the city of Vitoria-Gasteiz (the most intensive tests were carried our in the period July to October 2016). Two main functionalities of TDR were tested:

1. Its real-time responsiveness to queries for earliest-arrival-time (a.k.a. shortest-path) route plans of private cars.
2. The application of all the necessary updates of the traffic metadata kept in the UTKB, in order to incorporate the online traffic as recorded by the emergency reports generated by the Crowd-sourcing Module.

During the pilot execution, several earliest-arrival-time queries were submitted to the TDR cloud-service. The following data were recorded:

- For each query, we recorded the origin location, the destination location, the departure time from origin, the arrival time to destination, the distance, and for a medium-sized diesel car, the fuel consumption and CO_2 equivalent emissions.
- For each emergency report, we recorded the emergency report id, the start point and end point of the affected road segment, the distance of the road segment, the old and new travel time traversing the road segment and the start and end time point as the duration of the new travel time update.
- For each query before and after the absorption of online traffic updates, we recorded the emergency report id, the origin location, the destination location, the departure time from origin, the arrival time to destination, the distance, and for a medium-sized diesel car, the fuel consumption and CO_2 equivalent emissions.

Our pilot consolidation results revealed that more than 70% of the users were very satisfied with the TDR service, while another 17% were satisfied. Most of the users found the TDR service useful and easy to use without encountering any technical problems.

Figure 5 illustrates the procedure of an earliest-arrival-time query before and after an emergency report.

6   Conclusions

We presented the architecture of a cloud-based TDR service, aiming at providing fast real-time responses to arbitrary earliest-arrival-time queries as well as at updating efficiently the various traffic metadata kept in an urban traffic knowledge base, so that the service remains live-traffic aware. We also provided the implementation details of the core algorithmic TD routing engine, which is the most crucial module regarding the performance of the TDR service.

We plan to further enhance our cloud-based TDR service, in particular its core algorithmic TD engine, with the more sophisticated *hierarchical* algorithmic approaches presented recently in [6], which are expected to boost further the query time.

A   Hueristics for Improving Performance

We describe some heuristic algorithmic improvements, as well as some implementation details that we apply during the creation and maintenance of traffic metadata in the UTKB, in order to obtain even better performance, both wrt the required preprocessing space and wrt the efficiency of the query phase.

Approximately constant functions. The TRAP approximation method introduces one intermediate breakpoint per interval that satisfies the required approximation guarantee. To keep our algorithm space-efficient, the first thing that we do, when dealing with every subinterval of the time period, is that we check for approximately constant functions. More precisely, we perform an additional sampling at the middle point t_m of each $[t_s, t_f)$ and we consider the upper-approximation $\overline{\Delta}[\ell, v][t_s, t_f)$ to be "almost constant", if the following condition holds: $D[\ell, v](t_s) = D[\ell, v](t_f) = D[\ell, v](t_m)$. For those approximately constant functions, the insertion of the additional breakpoint at t_m is unnecessary.

Piecewise composition. Many shortest paths are likely to contain at least one arc with piecewise travel time, making the shortest-path function also piecewise. In our case, keeping the predecessor vertex of v in every sample of $\overline{\Delta}[\ell, v](t)$, allows us to analyze any ℓ-v approximate shortest path into two sub-paths ℓ-p-v. Starting from the destination v, we travel back to a predecessor p, as long as all vertices u up to p have constantly the same predecessor kept in the corresponding samples of $\overline{\Delta}[\ell, u](t)$, and all arcs involved in the p-v subpath have a constant travel-time function. Therefore, there is no need to keep any samples for the approximate function $\overline{\Delta}[\ell, v](t)$; instead, we store the necessary information in the form: (*constant-travel-time, predecessor-p, approximate-path-predecessor*). In this way, the approximate travel-time function $\overline{\Delta}[\ell, v](t)$ is given by taking into account $\overline{\Delta}[p, v](t)$ and the (approximately) constant travel-time from v to p, i.e., $\overline{\Delta}[\ell, v](t) = \overline{\Delta}[\ell, p](t) + \overline{\Delta}[p, v](t)$. In our experiments, this method leads to around 40–50% reduction of the space requirements.

Delay shifts. In cases that such a predecessor p does not exist, we have to store the approximate travel-time function as a sequence of breakpoints. However, the samples collected for $\overline{\Delta}[\ell, v](t)$ may have small delay variation. Based on this fact, the required space can be further reduced. We store the minimum travel-time value and for each leg we only need to store the small shift from this value. This conversion leads to around 5–10% reduction of the required preprocessing space.

Fixed range. For a one-day time period, departure-times have a bounded value range. The same holds for travel-times which are at most one-day for any query within a country or city area. When the considered precision of the traffic data is within seconds, we handle time-values as integers in the range $[0, 86399]$, rather than real values, sacrificing precision for space reduction. In particular, we convert all floating-point time-values t_f to integers t_i with fewer bytes and a given unit of measure. For a unit of measure s, the resulting integer is $t_i = \lceil t_f/s \rceil$ and needs size $\lceil \log_2(t_f/s)/8 \rceil$ bytes. Converting t_f to t_i results to an absolute error of at most 2 s. Therefore, for storing the time-values of approximate travel-time summaries, we can consider different resolutions, depending on the scale factor s, to achieve further reduction of the preprocessing space.

Compression. Since there is no need for all landmarks to be concurrently active, we can compress their data blocks. This method leads to significant reduction of the space requirements, especially for large-scale networks.

Contraction. The space of the preprocessed travel-time summaries can be further reduced if we consider a subset of vertices in the network as inactive. More precisely, we can conduct a preprocessing of the instance that contracts all vertices which are not junctions, i.e., they form paths with no intersections. Each such path can be represented by a single shortcut arc, which is added at the endpoints of the chain and equipped with an arc-traversal time function equal to the corresponding (exact) path-travel-time function. The arcs involved in the contracted paths are also considered as inactive. All contracted nodes are ignored and therefore the number of possible destinations from a landmark is smaller. At the query phase, these paths can be easily retrieved, by exploiting the appropriate information kept on all inserted shortcuts and all contracted nodes for this purpose.

Parallelism. We can speed up the preprocessing time for computing the one-to-all approximate travel-time functions, from properly selected landmarks towards all reachable destinations, as well as the real-time responsiveness to live-traffic reports, i.e., the re-computation of the travel-time summaries for the subset of affected landmarks, by exploiting the inherent parallelism of the entire process.

References

1. Agarwal, R., Godfrey, P.: Distance oracles for stretch less than 2. In: 24th Symposium on Discrete Algorithms (SODA 2013), pp. 526–538. ACM-SIAM (2013)
2. Bast, H., Delling, D., Goldberg, A., Müller-Hannemann, M., Pajor, T., Sanders, P., Wagner, D., Werneck, R.: Route planning in transportation networks. CoRR, abs/1504.05140 (2015)
3. Foschini, L., Hershberger, J., Suri, S.: On the complexity of time-dependent shortest paths. Algorithmica **68**(4), 1075–1097 (2014)
4. KaHIP - Karlsruhe High Quality Partitioning, May 2014
5. Kontogiannis, S., Michalopoulos, G., Papastavrou, G., Paraskevopoulos, A., Wagner, D., Zaroliagis, C.: Engineering oracles for time-dependent road networks. In: Algorithm Engineering and Experiments - ALENEX 2016 (SIAM 2016), pp. 1–14 (2016)
6. Kontogiannis, S., Wagner, D., Zaroliagis, C.: Hierarchical time-dependent oracles. In: Algorithms and Computation - ISAAC 2016. LIPICs, vol. 64, p. 47:1–47:13 (2015). http://arxiv.org/abs/1502.05222

7. Kontogiannis, S., Zaroliagis, C.: Distance oracles for time-dependent networks. Algorithmica **74**(4), 1404–1434 (2016)
8. MOVESMART project. http://www.movesmartfp7.eu/
9. Patrascu, M., Roditty, L.: Distance oracles beyond the Thorup-Zwick bound. SIAM J. Comput. **43**(1), 300–311 (2014)
10. Porat, E., Roditty, L.: Preprocess, set, query!. Algorithmica **67**(4), 516–528 (2013)
11. Sommer, C.: Shortest-path queries in static networks. ACM Comput. Surv. **46**, 1–31 (2014)
12. Sommer, C., Verbin, E., Yu, W.: Distance oracles for sparse graphs. In: 50th Symposium on Foundations of Computer Science (FOCS 2009), pp. 703–712. IEEE (2009)
13. Thorup, M., Zwick, U.: Approximate distance oracles. J. ACM **52**(1), 1–24 (2005)
14. Wulff-Nilsen, C.: Approximate distance oracles with improved preprocessing time. In: 23rd Symposium on Discrete Algorithms (SODA 2012), pp. 202–208. ACM-SIAM (2012)
15. Wulff-Nilsen, C.: Approximate distance oracles with improved query time. In: 24th Symposium on Discrete Algorithms (SODA 2013), pp. 539–549. ACM-SIAM (2013)

A Big Data Architecture for Traffic Forecasting Using Multi-Source Information

Yannis G. Petalas$^{(\boxtimes)}$, Ahmad Ammari, Panos Georgakis, and Chris Nwagboso

Sustainable Transport Research Group, Department of Civil Engineering,
Faculty of Science and Engineering,
University of Wolverhampton, Wolverhampton, UK
{I.Petalas,A.Ammari,P.Georgakis,C.Nwagboso}@wlv.ac.uk

Abstract. An important strand of predictive analytics for transport related applications is traffic forecasting. Accurate approximations of the state of transport networks in short, medium or long-term future horizons can be used for supporting traveller information, or traffic management systems. Traffic forecasting has been the focus of many researchers over the last two decades. Most of the existing works, focus on single point, corridor, or intersection based predictions with limited efforts to solutions that cover large metropolitan areas. In this work, an open big-data architecture for road traffic prediction in large metropolitan areas is proposed. The functional characteristics of the architecture, that allows processing of data from various sources, such as urban and inter-urban traffic data streams and social media, is investigated. Furthermore, its conceptual design using state-of-the-art computing technologies is realised.

Keywords: Big data · Intelligent transportation systems · Twitter · Social media · Natural Language Processing · Forecasting models

1 Introduction

Traffic forecasting plays a significant role in the transportation domain. Precise and accurate predictions of road traffic can improve the information services to travelers, lead to more efficient traffic information systems and improve road safety by reduction of accidents. Forecasting of the state of transport networks can result in the provision of mobility services that can reduce congestion, travel costs and CO_2 emissions, as well as lead to more efficient control and planning of traffic infrastructure.

Until now many models have been employed for traffic prediction [1], however the majority of them are applied and evaluated in an offline, rather static mode. Furthermore, their inputs are unidimensional and limited to single-point loop detectors, or clusters of detectors in close proximity. Thus, they lack the potential of utilizing dynamic information from other parts of the transportation network, or from other types of data sources.

Initial methodologies used for traffic forecasting were based on statistical analysis, and were followed by computational intelligence-based approaches for

© Springer International Publishing AG 2017
T. Sellis and K. Oikonomou (Eds.): ALGOCLOUD 2016, LNCS 10230, pp. 65–83, 2017.
DOI: 10.1007/978-3-319-57045-7_5

handling more efficient complex traffic conditions. The existing methods belong to three broad categories, parametric, non-parametric and hybrid and a review of the most prominent ones is presented in the next section.

It is believed that nowadays, traffic forecasting can benefit from a distributed big data architecture [2] that utilises data from multiple heterogeneous sources. This paper proposes an open big-data architecture that has the potential to meet this emerging need. Big data applications are characterized by the use of high volume, high variety and high velocity data. The proposed solution consists of a suite of forecasting models and a social media data mining component that processes data from twitter and relates them to the transport domain. Therefore, in this study high volume data in a variety of formats are retrieved in real-time from urban and inter-urban clusters of loop detectors, as well as through a stream of twitter feeds within the area of interest.

The paper is organized as follows. In Sect. 2, a review of forecasting models and uses of twitter data in the transport domain is presented. The data requirements for the operation of the proposed architecture is shown in Sect. 3. The models used for traffic forecasting are presented in Sect. 4, followed by a data mining approach for tweets in Sect. 5. Section 6 outputs the results of experimentation with fused data from traffic sensors and social media. The physical instantiation of the application, composed of a number of components and a distributed system implementation is discussed in Sect. 7. Section 8 concludes the paper with a reflective narration for the study.

2   Literature Review

2.1   Traffic Forecasting Models

Parametric Methodologies. This category involves statistical methodologies under the field of time series forecasting. An Autoregressive Integrated Moving Average (ARIMA) model was used for the short-term prediction of freeway traffic in the early 70s [3]. Since this initial effort, different variants of ARIMA were proposed to improve the prediction accuracy. These included, Kohonen ARIMA (KARIMA) [4], subset ARIMA [5], ARIMA with explanatory variables (ARIMAX) [6], spacetime ARIMA [7] and seasonal ARIMA (SARIMA) [8] to better handle the seasonality of the traffic data. Another set of parametric methodologies exploited Kalman Filtering (KF) and state space models. KF was used in [9] for incorporating an extra dimension (spatial) as part of a time-series model. Flow input from successive sensors on a transport corridor formed a multivariate time series state space model, which resulted to better predictions compared to that of a univariate ARIMA model. An enhanced time-series analysis integrated spatial characteristics and data as part of an ARIMA model [10]. Such state space models offer a multivariate approach, which makes KF methodologies suitable for network wide predictions [11]. In regards to other multivariate methodologies, a structural time-series (MST) technique, using the seemingly unrelated time-series equation (SUTSE), modeled time-series from flow observations in a network of junctions within a congested urban environment [12].

Non Parametric Methodologies. A limitation of the parametric methods presented above is the deterioration of their prediction accuracy when the data lack linearity. Since traffic related parameters exhibit stochastic and nonlinear characteristics the application of non-parametric methods in the field of traffic prediction gained momentum among researchers. The k-Nearest Neighborhood (k-NN) method was used for short-term freeway traffic forecasting and it was argued that it performed comparably with but not better than the linear time-series approach [13]. A dynamic multi interval traffic volume prediction model based on the k-NN non-parametric regression can be found in [14], while functional estimation techniques were applied as part of a kernel smoother for the auto-regression function in [15]. A local linear regression model and a Bayesian network approach for traffic flow forecasting were discussed in [16] and [17] respectively. Finally, an online learning weighted Support Vector Regression (SVR) model was presented in [18].

The majority of work in non-parametric methods is related to Artificial Neural Networks (ANN). Prediction outputs include Annual Average Daily Traffic (AADT) [19,20], flow, speed and occupancy [21,22], as well as mean travel times [23]. The State Space Neural Network (SSNN) is considered as a variant of Elman Neural Network and has been applied to predict urban travel time [24–27]. The result demonstrates that the SSNN is superior to other prevailing algorithms in terms of accuracy [27]. A Long Short-Term Neural Network (LSTM) was used to predict travel speed and outperformed the other recurrent neural networks (SSNN, Time Delay Neural Network) [28].

Hybrid Systems. There are many researchers that combined ANNs with other methodologies like optimization and Fuzzy Logic to improve the performance of ANNs. A Genetic Algorithm (GA) has been used for optimising the structure and hyperparameters of an ANN for predicting flows and travel times in urban networks [29,30]. Chaos Optimisation techniques have improved the performance of a GA supported ANN in [31]. A layered forecasting algorithm using ANN and Self-Organizing Maps (SOM) can be found in [32], while SOM has been adopted for the classification of traffic data as input to an ANN in [33]. In the former study, the layered structure was used to effectively support the forecasting task in situations where external events (e.g. incidents, social events, etc.) affected greatly recurrent traffic patterns.

A fuzzy neural model which classified input data using fuzzy rules and mapped input to output data using an ANN has been developed in [34], while a fuzzy rule base system that combined ANN and KF as part of a short-term forecasting model is presented in [35]. A deep-learning based coefficients optimization algorithm based on fuzzy ANN has been devised in [36]. Finally an aggregation approach for traffic flow prediction based on the Moving Average (MA), Exponential Smoothing (ETS), ARIMA, and ANN models has been proposed in [37].

2.2 Applications of Twitter Data in Transportation

More recently, there have been few emerging works on the exploitation of Intelligent Transport Systems (ITS) related information embedded in user-generated content of social data resources to support predictive analytics in traffic domain. Linear regression models have been employed for freeway traffic speed prediction using twitter data, weather and traffic information [38]. Twitter data had a relatively high sensitivity for predicting inclement weather (i.e., snow) during daytime. Twitter-based weather variables improved the predictive accuracy of the forecasting models. Traffic incident clustering and prediction was used for the development of a traffic management dashboard application in [39]. Tweets were classified to either insightful, or irrelevant to improve the system's performance in terms of traffic insights and alerts provision. Traffic incident management and detection on Dutch highways has been studied in [40]. In addition, spatio-temporal clusters of tweets on roads supported congestion detection in a study of Australian cities [41].

Prediction of traffic flow during sport events using a variety of methodologies like auto-regressive models, ANNs, SVRs and k-NN models has been examined in [42]. The authors reported that model enrichment with Twitter-engineered features improved prediction accuracy of regression models. An agent-based architecture for the detection of the movement of general public has been shown in [43]. Based on a city simulation framework, agent-based broadcasting of events extracted from tweets was claimed to be a source of human travel information useful for predicting traffic flow. Finally, twitter-engineered features extracted by KF models and semantic analysis of tweets were found to be able to overcome data processing limitations in existing baseline models, thus improving the accuracy of bus arrival time predictions [44].

3 Data Sources

The data sources that have been used for informing the development of the specifications for the proposed architecture are offered by Highways England [45], Birmingham City [46] and through Twitter's streaming APIs [47].

3.1 Highways England

Highways England, through its National Traffic Information Service (NTIS) initiative, provides real time traffic data (e.g. flow, average speed, travel time, headway, occupancy) to subscribed organisations. The format used is an extension of DATEX II, which is the European standard for traffic information exchange. For the purposes of this study, data from the following type of sensors, located on motorways enclosing the city of Birmingham, are being received:

1. "ANPR Journey Time Data": raw (unprocessed) real-time Automatic Number Plate Recognition (ANPR) journey time data, measured between ANPR camera sites.

2. "MIDAS Gold Loop Traffic Data": raw (unprocessed) real-time traffic data measured by Motorway Incident Detection and Automatic Signaling (MIDAS) loop sensors.
3. "Fused Sensor-only PTD": real-time Processed Traffic Data (PTD), calculated from raw sensor traffic data.
4. "TMU Loop Traffic Data": raw (unprocessed) real-time traffic data measured by Traffic Monitoring Unit (TMU) loop sensors.

Data from MIDAS and fused sensors are being pushed to subscribers every one minute, while data from ANPR and TMU systems every five minutes.

3.2 Birmingham City Council

Birmingham City Council provides traffic data through their open data infrastructure. REST endpoint calls are used to collect the traffic data in a proprietary format. The provided data are pulled every 5 to 10 min and include flows, average speeds and travel times, from loop detectors and ANPR installations.

3.3 Twitter

The Twitter APIs provide access to four main objects that have been considered for this work. These include:

1. "Tweets": The basic atomic building block of all things. Tweets are also known as status updates. Tweets can be embedded, replied to, liked, unliked and deleted. The key tweet fields streamed/retrieved for analysis are coordinates, creation timestamp, tweet id, language, place, and text.
2. "Users": Users can be anyone or anything. They tweet, follow, create lists, have a home timeline, can be mentioned, and can be looked up in bulk. The key user fields considered are account creation timestamp, description, number of followers, number of friends, whether the account has enabled the possibility of geotagging, account id, language, location, and number of tweets issued by the user.
3. "Entities": Provide metadata and additional contextual information about content posted on Twitter. Entities are never divorced from the content they describe. Entities are returned wherever tweets are found in the API. Entities are instrumental in resolving URLs. The key entity keys considered are hashtags and user mentions.
4. "Places": Specific, named locations with corresponding geo-coordinates. They can be attached to Tweets by specifying a place id when tweeting. Tweets associated with places are not necessarily issued from that location but could also potentially be about that location. Places can be searched for and Tweets can be requested by place id. Considered place fields are attributes, bounding box of coordinates which encloses this place, country, country code, human-readable representation of the places name (e.g. Birmingham), type of location represented by this place (e.g. city).

The volume, velocity and variety of the information received in real time by the aforementioned sources can be seen in Table 1.

Table 1. Volume and velocity for the available data sources in the region of Birmingham city

Data source	Volume	Velocity	Variety
Tweets	300 MB per day	1 min	Twitter objects in json format
Urban traffic	600 MB per day	5–10 min	Proprietary json documents
Motorway traffic	3 GB per day	1–5 min	Datex II compliant XML messages

4 Forecasting Models

Before using these data as input to the forecasting models, a pre-processing procedure takes place to enhance their quality. This procedure involves the following steps:

- *Data cleaning*, which involves detecting samples with values that do not lie in a specific range and removing duplicates. For example, traffic measurements with many zero values are being discarded.
- *Computation of the percent of missing values per day*, which if is above a threshold (very high) the respective days are removed from the dataset.
- *Examination for big plateau*, where if a lot of measurements with the same value for consecutive time periods are received, is an indication of invalidity for the specific set of data. For example, consecutive sensor readings of 80 km/h for average speed over a period of many hours.
- *Imputation of missing values*, when valid measurements for the whole duration that the loop detectors operate (e.g. measurements per minute) are not possible. There are cases where measurements are not available due to malfunction of the loop detectors, or a possible break down in the data infrastructure (e.g. database). There are different strategies to handle missing values, such as to use the mean historical value of the available measurements, or their most frequent value. In this study linear interpolation has been used to replace the missing values.
- *Aggregation of the data* that are being received in various frequencies. Aggregation in different time scales e.g. per 5, 10, 15 min has to be adopted before the application of the forecasting models.

After the pre-processing procedures, data with improved quality can be employed from the forecasting models. Various models have been used until now for traffic forecasting, ranging from statistics and computational intelligence as described in Sect. 2. A number of such models have been incorporated in the proposed architecture in order to cover this wide range. From statistical sciences the methods considered include Auto-Regressive Integrated Moving Average (ARIMA) [48], Exponential Smoothing models (ETS) [49], Dynamic linear models [50,51], Linear Regression [52] and Random forest Regression [53]. From computational intelligence, Artificial Neural Networks (ANNs) [54] and Support Vector Regression (SVR) [55,56] have been used. Regarding ARIMA and ETS, packages from R project [57] were explored, e.g. the forecasting package [58]. From DLM, the two basic models, the local level and the local linear trend were applied, using Kalman Filtering for the estimation of the models [59]. Regarding ANNs, Time Delay Neural Networks (TDNN) [54] were used in the experiments, since they are appropriate for applications with sequential data. They have the same structure as multilayer feed forward neural networks with a delay tape of observations. If the predicted value is $x(n)$, the inputs of the TDNN will be the s previous values $x(n - 1), x(n - 2), \ldots, x(n - s)$. The ANN model was implemented in Python using the Keras library [60]. Keras is a library that allows the development of deep learning applications and it includes a suite of training methodologies (optimisers) that are suitable for online/real time learning. For linear regression, Random Forest Regression and Support Vector Regression the implementation in Python was supported by the scikit-learn library [61].

There are two phases related to the development of forecasting models. The first deals with the training of each model and it usually takes place offline, while the second phase is where the trained models receive their required input and provide predictions. For some models like DLM, ANNs and SVR the two phases may overlap such that they can be updated per sample of data. Regarding the training phase, requests are made to the storage of historical data, which may span across a period of months. For the statistical models (ARIMA, ETS, DLM) the used dataset is split in two parts, the training subset (80%) and the testing subset (20%). The fit of the models takes place using the training set and their evaluation using the testing set. Well known measures in the literature such as Mean Absolute Percentage Error (MAPE), Mean Absolute Error (MAE) and Root Mean Square Error (RMSE) are used for the evaluation of the models [49].

For the rest of the forecasting models, which are primarily regression models, the dataset is split in 3 parts. Training set 60%, validation set 20% and testing set 20%. The validation subset is used for the computation of parameters specific to each method. These are, in random forest the number of trees/estimators, in SVR the type of kernel and its associate parameters and in ANNs the number of hidden layers. The final evaluation takes place with the usage of the testing set and the measures that have been mentioned above (MAPE, MAE, RMSE). The output of these models is the prediction for the next time step, while the predictors are recent values (in this study 2 previous observations) from traffic data, the mean historical value per hour and minute and other variables such as the current day, hour and minute. Additional predictors could use information generated from tweets related to traffic conditions.

There are two basic approaches for multistep ahead prediction. The first one is the incremental, where one step ahead prediction is made and the result is fed back as an input to the model for the next step prediction. This process is repeated until the number of desired steps ahead is reached. The second one is to use a separate model for each step ahead that needs to be predicted. For

ANNs there is a third option to have as outputs the number of the steps that we want to predict ahead. If we want to predict 12 steps ahead, the ANN will have 12 neurons at the output layer. Regarding ANNs this was found to be the best approach and was adopted in the experiments. For the other regression models the second approach was used, e.g. building a separate model for each step ahead.

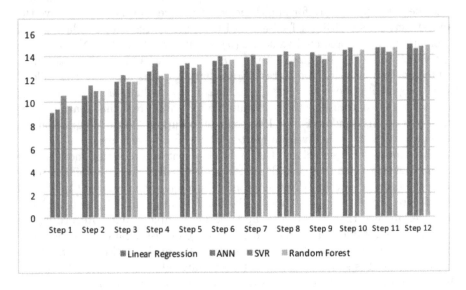

Fig. 1. MAPE 12 step ahead flow forecasting for a motorway sensor (M6)

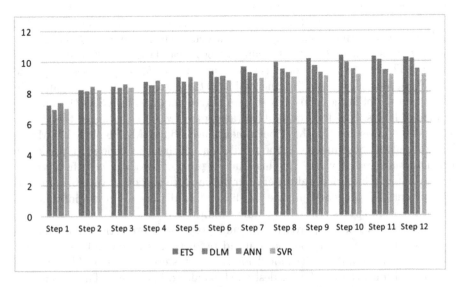

Fig. 2. MAPE 12 step ahead average speed forecasting for an urban sensor (N30162Y)

In Figs. 1 and 2 some experimental results are presented. Traffic flow forecasting per 1 h ahead (12 steps) for one sensor located on a motorway (M6)
and average speed forecasting per 1 h ahead for one sensor located close to the
city of Birmingham. It can be noticed that the error increases for longer steps
ahead and that per step of prediction the model with the best accuracy may be
different.

5   Social Information Miner

Information generated by social media, and in particular from Twitter, can
support traffic prediction applications in locations where sensors are sparsely
located. The social media data mining approach used in this study employs different text pre-processing, Natural Language Processing (NLP), text mining, and
text classification algorithms for the realisation of the sub-components presented
in Fig. 3. A brief description of each sub-component can be found below:

- Two data streaming adapters have been developed to collect tweets from the
  public Twitter Streaming API. The first is using a list of traffic informing
  twitter accounts, while the second is receiving public tweets using a geolocation filter.
- The text processing engine adopts NLP mechanisms to pre-process and
  clean the tweets received from streaming API. The pre-processing pipeline
  is depicted in Fig. 5.
- The relevance filtration classifier identifies tweets that are related to the transport domain. Different machine learning techniques and N-gram range sets

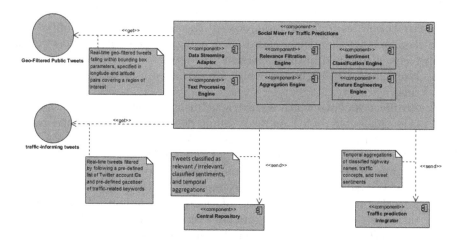

Fig. 3. Social media miner architecture

have been trialed for generating a classification model with high accuracy percentages. In more particular, Multinomial Naïve Bayes (MNB), Support Vector Classification (SVC), Random Forest (RF), Artificial Neural Networks (ANN) were used, and the derived results can be seen in Table 2.

- A Sentiment classifier has been developed for identifying the polarity (positive, neutral, negative) of pre-processed tweets. The classifier enriches tweets with sentiment terms (good, bad, excellent, etc.) based on a mapping gazetteer that maps the traffic-related terms to the sentiment terms that will bias the overall sentiment scores of the tweets as obtained from a lexical database. For example, the tweet *"All lanes are now open on A21 Southbound btwn A25 near Chipstead and A225 near Sevenoaks following an earlier collision"* will be enriched with sentiment terms according to the mapping:

"now open" $\longrightarrow$ **"excellent great"**, **"collision"** $\longrightarrow$ **"bad"**

Following enrichment, the tweet is fed into the SentiWordNet lexical resource for opinion mining [62] to determine its overall sentiment score. The boxplots depicted in Fig. 4 illustrate the median sentiments scores of the tweets that include traffic-related terms in the training dataset. There is an obvious discrepancy of sentiment median scores between terms implying positive context (e.g. open, moving, cleared, released) and terms implying negative context (e.g. accident, congestion, flooding, delays), whereas terms implying neutral context (e.g. traffic, management) are mentioned in tweets with median sentiment scores around zero.

- Aggregation engine to produce time series aggregations of traffic mentions and sentiment polarities from the pre-processed tweets.
- Feature engineering sub-component to derive extended features from the pre-processed tweets, such as extracting the named entities (people, organizations, locations). The objective of the feature engineering sub-component is to extract entities from the pre-processed tweets. The extracted entities can be classified into three main classes, depending on the entity type as well as the technique(s) used in the extraction process:

1. **Names of UK highways** which are extracted using regular expression rules.
2. **Traffic concepts** which are extracted using a traffic gazetteer that has been generated based on text summaries and analysis of the tweets received from the traffic informing accounts and which includes word frequency, collocation, and concordance analysis.
3. **Names of recognized locations, organizations, and people**, which are extracted using the Stanford Named Entity Recognition (NER) recognizer. For the purposes of mining information for supporting traffic predictions the general pre-trained Stanford recognizers for *PERSON, ORGANIZATION* and *LOCATION* entities has been used.

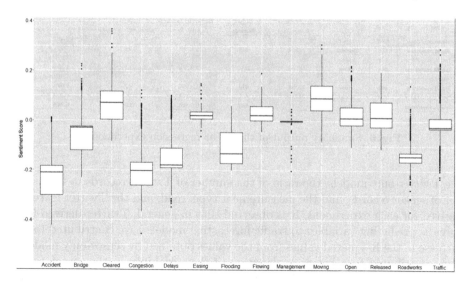

Fig. 4. Comparison between the sentiment scores of 15 traffic-related terms

Fig. 5. Text processing and cleaning pipeline

Table 2. Confusion matrix for relevant (positive) and noise (negative) testing tweets with varying classification models

Algorithm	Actual class	Predicted negative	Predicted positive	Class precision	Class recall	Class F1 score
MNB	Negative	TN: 2,370	FP: 27	99.04%	98.87%	98.95%
	Positive	FN: 23	TP: 2,580	98.96%	99.12%	99.04%
SVC	Negative	TN: 2,378	FP: 19	99.08%	99.21%	99.14%
	Positive	FN: 22	TP: 2,581	99.27%	99.15%	99.21%
RF	Negative	TN: 2,307	FP: 90	80.95%	96.25%	87.94%
	Positive	FN: 543	TP: 2,060	95.81%	79.14%	86.68%
ANN	Negative	TN: 2,324	FP: 73	98.64%	96.95%	97.79%
	Positive	FN: 32	TP: 2,571	97.24%	98.77%	98.00%

6 Data Fusion Experimentation

Following the development of the forecasting models and the social miner component, linear regression experiments were performed for investigating the prediction potential when information from heterogeneous sources is fused. Figure 6 summarizes the outputs from the experiments performed. In particular, the table highlights the percentage of the improvement of R-squared values after adding the Twitter features as predictors to the sensor-only models, the R-squared of

Twitter Feature	Road	Measurement Type	No of Sensor Readings	No of Twitter Records	Twitter-to-Sensor Records Ratio	R-squared of Twitter-only model	R-squared Improvement from Sensor-only model
Road mentions	M42	Speed	28246	363	1.285%	0.16	7.46%
Road mentions	M6	Speed	28254	473	1.674%	0.18	5.79%
Traffic mentions	M6	Speed	96780	2916	3.013%	0.06	1.87%
Traffic mentions	M6	Flow	96780	2916	3.013%	0.41	1.50%
Road mentions	M42	Flow	28246	363	1.285%	0.17	1.08%
Road mentions	M6	Flow	28254	473	1.674%	0.42	1.07%
Traffic mentions	M54	Speed	96759	488	0.504%	0.06	0.25%
Traffic mentions	M54	Flow	96759	488	0.504%	0.08	0.24%

Fig. 6. Summary output of the linear regression experiments

the Twitter-only models, the ratio of the number of Twitter records to the number of sensor records, and the measurement type, road, and the Twitter features derived for each experiment. It is observed that in general, Twitter-derived features as predicting variables to traffic forecasting models have contributed to an increase in the R-squared values from the values obtained by sensor-only models, thus contributed to a better explanation of the variability in the sensor readings as a response variable. However, the summary shows that the significant improvements occur, firstly, when the measurement type is the average speeds of the vehicles rather than the vehicles flows, and secondly, when using road mentions, rather than traffic mentions, as Twitter-derived features. The range of the absolute differences of the R-squared values between the Twitter-aided models and the sensor-only models is between 0.038 and 0.032 in the maximum two increased cases of 7.46% and 5.79%, and 0.002 in the minimum case of 0.24%. Although these differences are relatively small, they are not very far from the differences obtained by the analysis of similar studies, such as the study of [38], where the absolute differences of the R-squared values for one of the tried datasets (NFIA) between some Twitter and No-Twitter models developed to predict average speeds in inclement weather situations were between 0.05 and 0.07 (Daytime), and between 0.01 and 0.02 (Night-time).

On the other hand, using Twitter-derived features alone as explanatory attributes to the variability in the sensor readings, whether these features are extracted from the road mentions or the traffic-related concepts, provides the best R-squared results when the measurement type is the vehicles flows, rather than the average speeds. It is also noticed that having too insignificant Twitter-to-sensor records ratios (e.g. less than 1%) due to relatively low volumes of tweets about traffic events in particular roads, such as M54, is correlated with obtaining a poor explanation of the variability in the sensor readings. This applies on the two measurements types (flows and speeds) and when Twitter-derived features are used in both Twitter-only models and Twitter-and-sensor models. This motivates the utilization of more techniques, such as the use of social network analysis (SNA), to find more traffic-related tweets by discovering further Twitter accounts that post tweets about traffic events and road status.

7 Big Data Architecture

In this section, the proposed distributed architecture for traffic forecasting, as shown in Fig. 7, will be described. Any sustainable architecture should be modular, extensible, fault-tolerant and be able to satisfy the specifications related to traffic forecasting as these change following technological developments.

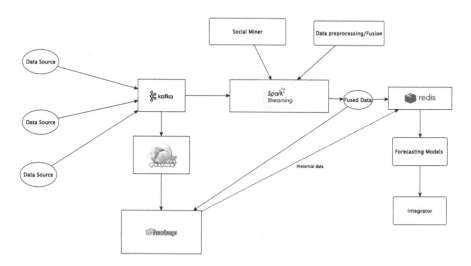

Fig. 7. Distributed architecture for traffic forecasting

The inputs to the system are received through various available data sources and are being ingested in Apache Kafka [63]. Apache Kafka is an open source, distributed publish-subscribe (pub-sub) messaging system. Its main characteristics are persistence and replication messaging, high throughput, low latency, fast real time consumption of messages, high availability and no loss of data.

In a publish-subscribe system, messages persisted to a topic are produced by **publishers** and consumed by **subscribers**. In this study, each available data source is a publisher and can produce data to a topic in Kafka. Therefore, there will be a topic for traffic data from Highways England, another topic for traffic data from Birmingham City Council and a third topic for Twitter data. One can notice that a new data source can be easily integrated in this architecture. This will require the creation of a new topic in Kafka and from there data can be consumed from the other components of the architecture. With the usage of a pub-sub system, like Kafka, the decoupling of the data sources from the other parts of the architecture is achieved. From Kafka, many different pipelines can be initiated towards other components of the architecture.

As mentioned in Sect. 4, the first requirement for traffic forecasting is the training of the models using historical data, while the second is the performance of predictions using the trained models. For the second requirement, real time data are needed to serve as input to the forecasting models. Thus, in order to fulfill these requirements two pipelines/flows in the architecture are proposed. In the first pipeline, the data from Kafka are transferred to Hadoop where they can be stored for lengthy periods of time. The volume of data per year can reach terabytes, thus the usage of the Hadoop ecosystem [64] is proposed. Hadoop allows the storage of data on a distributed file system (HDFS) that can be expanded by the integration of additional commodity hardware.

The data can reach Hadoop through Flume which can store them to HDFS, or to another storage supported in Hadoop. For the proposed platform, HBase, which is a distributed, column-oriented database that uses HDFS for its underlying storage is the preferred solution. This is due to its ability to store and retrieve structured, or semi-structured data using random access mechanisms [65]. Apache Flume [66] is a distributed, reliable, and available system for the efficient collection, aggregation, and movement of streaming data. Flume is more suitable for data ingestion in Hadoop. It has very easy configuration and can handle many common issues in writing data to HDFS or HBase, such as reliability, optimal file sizes, file formats, updating metadata, and partitioning. Kafka and Flume can complement each other and often the synergy of the two modules appear in big data architectures [67]. If besides traffic forecasting, it is desired to provide analytics/insights related to the archived traffic data, Apache Hive could be used to build a distributed data warehouse, along with a visualization/dashboard tool. Hive manages data stored in HDFS, or HBase and provides a query language for performing MapReduce operations in easier and friendlier ways.

In the second pipeline/flow, Spark Streaming [68] is proposed to receive the streaming data from Kafka and serve them to the forecasting models as input. Spark Streaming is Sparks module for handling data as soon as they arrive from various data sources. It is a very similar API to batch processing jobs, and thus code written for batch jobs can be reused. Spark Streaming receives data streams every n seconds in a structure called DStream or discretized stream. Internally, each DStream is represented as a sequence of Resilient Distributed datasets (RDDs which are the primary abstraction in Spark), arriving at each time step. Thus, the APIs that Spark provides (Spark SQL or Spark MLib), can be incorporated in Spark Streaming. Spark Streaming uses a "micro-batch" architecture. This allows the streaming computation to be treated as a continuous series of batch computations on small batches of data. Since the data will not be used immediately (milliseconds after they are available) from the forecasting models, but be aggregated per 5 minutes or more, the "micro-batch" architecture is not a limitation for such a case and the dynamic load balancing and high throughput that Spark Streaming provides can be advantageous [69].

A DStream can be created for each topic in Kafka which, as previously mentioned, is associated with each of the available data sources. The window feature of Spark Streaming, provides the capability to perform cleaning, aggregation and enrichment to the received data. A window contains many batches of data and could intuitively have data from the last 5–15 minutes or more. For example, related to the traffic data sources, the pre-processing that was described in Sect. 4 can take place during the period that a window lasts.

Transformations of DStreams, like joins, can be accomplished to align and fuse the different DStreams in order to create a common dataset that will be used from the forecasting models. This dataset could have in each row the following information sensorId, datetime, traffic measurements (flow, average speed), number of tweets related to accidents/congestion/delay, visibility (weather info), precipitation (weather info). This fused dataset from Spark Streaming can be stored to the Hadoop/HBase to be used for the offline training of forecasting models and to an in memory database for providing real time input to the trained forecasting models. The role of the in memory database is to serve as a cache, storing the most recent data, thus allowing for rapid retrieval and real-time processing. Redis [70], which is an open source (BSD licensed), in-memory data structure store, used as a database, cache and message broker is proposed for this task. In the presented case, Redis will be used as in memory database in order to support predictions in near real time. Redis has data structures such as, sorted set and hash, for handling efficiently time series and a connector is available for communicating with Spark.

The last part of the architecture refers to the component that contains the forecasting models and the social media miner. This component is composed from a number of forecasting models as described in Sect. 4. The forecasting models can be trained using the data stored in Hadoop/HBase and make predictions using recent observations from the Redis database. In the case of the models that can work on-line, these models will be updated using the real time data from Redis. Similarly, the twitter processing pipeline presented in Sect. 5 can be fed with tweets from the same components of the architecture. An integrator component is proposed to receive forecasts from the set of models and use assemble techniques for providing a final more accurate prediction. The final output from the integrator can be used to feed external traveler information systems (for example route planners), or traffic management systems (for example traffic signal control systems).

8   Conclusions

Experimentations presented in this paper demonstrated that the performance of different forecasting models varies depending on the type of traffic data used, forecasting timesteps, time of the day and nature of the transport network that a sensor is located. Therefore, and to achieve optimal predictions, the use of a multiple of models must be available in a traffic forecasting application. Such availability results in further complexities due to the diverse requirements for training and evaluation of each model in real-time. In addition, the use of crowd sourcing data enrichment has the potential to improve the forecasting outcomes.

Due to the above, modern traffic forecasting requires real-time collection, processing and storage of high volume, variety and velocity data. In this paper, an open big data distributed architecture has been proposed for traffic forecasting.

It offers optimised data aggregation (cleaning, filtering, etc.), it allows integration and usage of real-time and historical data and enables cluster-based processing.

Acknowledgments. This research was supported by the European Union's Horizon 2020 research and innovation programme under grant agreement No. 636160-2, the Optimum project, http://www.optimumproject.eu.

References

1. Vlahogianni, E., Karlaftis, M., Golias, J.: Short-term traffic forecasting: where we are and where we're going. Transp. Res. Part C Emerg. Technol. **43**, 3–19 (2014)
2. Vlahogianni, E.I., Park, B.B., van Lint, J.W.C.: Big data in transportation and traffic engineering. Transp. Res. Part C Emerg. Technol. **58**, 161 (2015)
3. Ahmed, M.S., Cook, A.R.: Analysis of freeway traffic time-series data by using Box-Jenkins techniques. Transp. Res. Rec. **722**, 1–9 (1979)
4. VanderVoort, M., Dougherty, M., Watson, S.: Combining Kohonen maps with ARIMA time series models to forecast traffic flow. Transp. Res. C Emerg. Technol. **4**(5), 307–318 (1996)
5. Lee, S., Fambro, D.: Application of subset autoregressive integrated moving average model for short-term freeway traffic volume forecasting. Transp. Res. Rec. **1678**, 179–188 (1999)
6. Williams, M.: Multivariate vehicular traffic flow prediction evaluation of ARIMAX modeling. Transp. Res. Rec. **1776**, 194–200 (2001)
7. Kamarianakis, Y., Prastacos, P.: Forecasting traffic flow conditions in an urban network comparison of multivariate and univariate approaches. Transp. Res. Rec. **1857**, 74–84 (2003)
8. Williams, M., Hoel, L.A.: Modeling and forecasting vehicular traffic flow as a seasonal ARIMA process: theoretical basis and empirical results. J. Transp. Eng. **129**(6), 664–672 (2003)
9. Stathopoulos, A., Karlaftis, M.G.: A multivariate state space approach for urban traffic flow modeling and prediction. Transp. Res. C Emerg. Technol. **11**(2), 121–135 (2003)
10. Kamarianakis, Y., Prastacos, P.: Spacetime modeling of traffic flow. Comput. Geosci. **31**(2), 119–133 (2005)
11. Whittaker, J., Garside, S., Lindveld, K.: Tracking and predicting a network traffic process. Int. J. Forecast. **13**, 51–61 (1997)
12. Ghosh, B., Basu, B., O'Mahony, M.: Multivariate short-term traffic flow forecasting using time-series analysis. IEEE Trans. Intell. Transp. Syst. **10**(2), 246–254 (2009)
13. Davis, G.A., Nihan, N.L.: Nonparametric regression and short-term freeway traffic forecasting. J. Transp. Eng. **117**(2), 178–188 (1991)
14. Chang, H., Lee, Y., Yoon, B., Baek, S.: Dynamic near-term traffic flow prediction: system oriented approach based on past experiences. IET Intell. Transp. Syst. **6**(3), 292–305 (2012)
15. El Faouzi, N.E.: Nonparametric traffic flow prediction using kernel estimator. In: Proceedings of the 13th ISTTT, pp. 41–54 (1996)
16. Sun, H.Y., Liu, H.X., Xiao, H., He, R.R., Ran, B.: Use of local linear regression model for short-term traffic forecasting. Transp. Res. Rec. **1836**, 143–150 (2003)
17. Sun, S., Zhang, C., Guoqiang, Y.: A Bayesian network approach to traffic flow forecasting. IEEE Intell. Transp. Syst. Mag. **7**(1), 124–132 (2006)

18. Jeong, Y.S., Byon, Y.J., Castro-Neto, M.M., Easa, S.M.: Supervised weighting-online learning algorithm for short-term traffic flow prediction. IEEE Trans. Intell. Transp. Syst. **14**(4), 1700–1707 (2013)
19. Faghri, A., Hua, J.: Roadway seasonal classification using neural networks. J. Comput. Civ. Eng. **9**(3), 209–215 (1995)
20. Lingras, P., Adamo, M.: Average and peak traffic volumes: neural nets, regression, factor approaches. J. Comput. Civ. Eng. **10**(4), 300 (1996)
21. Dougherty, M.S., Cobbett, M.R.: Short-term inter-urban traffic forecasts using neural networks. Int. J. Forecast. **13**(1), 21–31 (1997)
22. Dia, H.: An object-oriented neural network approach to short-term traffic forecasting. Eur. J. Oper. Res. **131**(2), 253–261 (2001)
23. Li, R., Rose, G.: Incorporating uncertainty into short-term travel time predictions. Transp. Res. Part C Emerg. Technol. **19**(6), 1006–1018 (2011)
24. Van Lint, J.W.C., Hoogendoorn, S.P., Zuylen, H.V.: Freeway travel time prediction with state-spaced neural networks: modeling state-space dynamics with recurrent neural networks. Transp. Res. Rec. J. Transp. Res. Board **1811**, 30–39 (2002)
25. Van Lint, J.W.C.: Reliable real-time framework for short-term freeway travel time prediction. J. Transp. Eng. **130**(12), 921–932 (2004)
26. Van Lint, J.W.C., Hoogendoorn, S.P., Zuylen, H.V.: Accurate freeway travel time prediction with state-space neural networks under missing data. Transp. Res. Part C **13**, 347–369 (2005)
27. Liu, H., Zuylen, H.V., Lint, H.V., Salomons, M.: Predicting urban arterial travel time with state-space neural networks and Kalman filters. Transp. Res. Rec. J Transp. Res. Board **1968**, 99–108 (2006)
28. Ma, X., Tao, Z., Wang, Y., Yu, H., Wang, Y.: Long short-term memory neural network for traffic speed prediction using remote microwave sensor data. Transp. Res. Part C **54**, 187–197 (2015)
29. Vlahogianni, E.I., Karlaftis, M.G., Golias, J.C.: Optimized and meta-optimized neural networks for short-term traffic flow prediction: a genetic approach. Transp. Res. Part C **13**(3), 211–234 (2005)
30. Khosravi, A., Mazloumi, E., Nahavandi, S., Creighton, D., Van Lint, J.W.C.: A genetic algorithm-based method for improving quality of travel time prediction intervals. Transp. Res. Part C **19**(6), 1364–1376 (2011)
31. He, W., Lu, T., Wang, E.: A new method for traffic forecasting based on the data mining technology with artificial intelligent algorithms. Res. J. Appl. Sci. Eng. Technol. **5**(12), 3417–3422 (2013)
32. Zhu, J., Zhang, T.: A layered neural network competitive algorithm for short-term traffic forecasting. In: Computational Intelligence and Software Engineering, pp. 1–4 (2009)
33. Chen, H., Grant-Muller, S., Mussone, L., Montgomery, F.: A study of hybrid neural network approaches and the effects of missing data on traffic forecasting. Neural Comput. Appl. **10**(3), 277 (2001)
34. Yin, H.: Urban traffic flow prediction using a fuzzy-neural approach. Transp. Res. Part C Emerg. Technol. **10**, 85–98 (2002)
35. Dimitriou, L., Tsekeris, T., Stathopoulos, A.: Adaptive hybrid fuzzy rule-based system approach for modeling and predicting urban traffic flow. Transp. Res. C Emerg. Technol. **16**(5), 554–573 (2008)
36. Lu, H.P., Sun, Z.Y., Qu, W.C., Wang, L.: Real-time corrected traffic correlation model for traffic flow forecasting. Math. Problems Eng. **501**, 348036 (2015)

37. Tan, M.C., Wong, S.C., Xu, J.M., Guan, Z.R., Peng, Z.: An aggregation approach to short-term traffic flow prediction. IEEE Trans. Intell. Transp. Syst. **10**(1), 60–69 (2009)
38. Lin, L., Ni, M., He, Q., Gao, J., Sadek, A.W.: Modeling the impacts of inclement weather on freeway traffic speed: exploratory study with social media data. Transp. Res. Rec.: J. Transp. Res. Board **2482**, 82–89 (2015)
39. Tejaswin, P., Kumar, R., Gupta, S.: Tweeting traffic: analyzing twitter for generating real-time city traffic insights and predictions. In: Proceedings of the 2nd IKDD Conference on Data Sciences, p. 9 (2015)
40. Steur, R.J.: Twitter as a spatio-temporal source for incident management. Thesis, Utrecht University, Utrecht (2015)
41. Gong, Y., Deng, F., Sinnott, R.O.: Identification of (near) real-time traffic congestion in the cities of Australia through Twitter. In: Proceedings of the ACM First International Workshop on Understanding the City with Urban Informatics, pp. 7–12 (2015)
42. Ni, M., He, Q., Gao, J.: Using social media to predict traffic flow under special event conditions. In: The 93rd Annual Meeting of Transportation Research Board (2014)
43. Pathania, D., Karlapalem, K.: Social network driven traffic decongestion using near time forecasting. In: Proceedings of the 2015 International Conference on Autonomous Agents and Multiagent Systems, pp. 1761–1762 (2015)
44. Abidin, A.F., Kolberg, M., Hussain, A.: Integrating Twitter traffic information with Kalman filter models for public transportation vehicle arrival time prediction. In: Trovati, M., Hill, R., Anjum, A., Zhu, S.Y., Liu, L. (eds.) Big-Data Analytics and Cloud Computing, pp. 67–82. Springer, Cham (2015). doi:10.1007/978-3-319-25313-8_5
45. Highways England. http://www.highways.gov.uk/
46. Birmingham City Council Open Data. http://butc.opendata.onl/AL_OpenData
47. Twitter. https://dev.twitter.com/
48. Box, G.E.P., Jenkins, G.M.: Time Series Analysis: Forecasting and Control, 2nd edn. Holden-Day, San Francisco (1976)
49. Hydman, R., Athanasopoulos, G.: Forecasting Principles and Practice. OTexts, Melbourn (2013)
50. Petris, G., Petrone, S., Campagnoli, P.: Dynamic Linear Models with R. Springer, New York (2009)
51. Commander, J., Koopman, S.J.: An Introduction to State Space Time Series Analysis. Oxford University Press, Oxford (2007)
52. Hastie, T., Tibshirani, R., Friedman, J.: The Elements of Statistical Learning: Data mining, Inference and Prediction, 2nd edn. Springer, New York (2009)
53. Breiman, L.: Random forests. Mach. Learn. **45**(1), 5–32 (2001)
54. Haykin, S.: Neural Networks: A Comprehensive Foundation, 2nd edn. Pearson, Prentice Hall, Upper Saddle (1999)
55. Vapnik, V.N.: The Nature of Statistical Learning Theory. Springer, New York (1996)
56. Vapnik, V.N.: Statistical Learning Theory. Wiley, New York (1998)
57. R project. https://www.r-project.org
58. Hyndman, R.J., Yeasmin Khandakar, Y.: Automatic time series forecasting: the forecast package for R. J. Stat. Softw. **26**(3), 1–22 (2008)
59. Pykalman. https://pykalman.github.io
60. Keras Library. https://github.com/fchollet/keras

61. Pedregosa, F., Varoquaux, G., Gramfort, A., Michel, V., Thirion, B., Grisel, O., Blondel, M., Prettenhofer, P., Weiss, R., Dubourg, V., Vanderplas, J., Passos, A., Cournapeau, D., Brucher, M., Perrot, M., Duchesnay, E.: scikit-learn: machine learning in Python. J. Mach. Learn. Res. **12**, 2825–2830 (2011)

62. Esuli, A., Sebastiani, F.: SENTIWORDNET: A publicly available lexical resource for opinion mining. In: Proceedings of the 5th Conference on Language Resources and Evaluation (LREC 2006), pp 417–422 (2006)

63. Apache Kafka. https://kafka.apache.org/

64. Apache Hadoop. https://hadoop.apache.org

65. Dimiduk, N., Khurana, A.: HBase in Action. Manning, Shelter Island (2013)

66. Apache Flume. https://flume.apache.org

67. Grover, M., Malaska, T., Seidman, J., Shapira, G.: Hadoop Application Architectures. O'Reilly Media, Beijing (2015)

68. Apache Spark. https://spark.apache.org

69. Karau, H., Konwinski, A., Wendell, P., Zaharia, M.: Learning Spark. O'Reilly Media, Sebastopol (2015)

70. Redis. https://redis.io

Algorithmic Aspects of Large-Scale Data Stores

Mining Uncertain Graphs: An Overview

Vasileios Kassiano, Anastasios Gounaris, Apostolos N. Papadopoulos[(✉)],
and Kostas Tsichlas

Department of Informatics, Aristotle University of Thessaloniki, Thessaloniki, Greece
{vkassiano,gounaria,papadopo,tsichlas}@csd.auth.gr

Abstract. Graphs play an important role in modern world, due to their widespread use for modeling, representing and organizing linked data. Taking into consideration that most of the "killer" applications require a graph-based representation (e.g., the Web, social network management, protein-protein interaction networks), efficient query processing and analysis techniques are required, not only because these graphs are massive but also because the operations that must be supported are complex, requiring significant computational resources. In many cases, each graph edge e is annotated by a probability value $p(e)$, expressing its *existential uncertainty*. This means that with probability $p(e)$ the edge will be present in the graph and with probability $1 - p(e)$ the edge will be absent. This gives rise to the concept of *probabilistic graphs* (also known as *uncertain graphs*). Formally, a probabilistic graph $\mathcal{G}$ is a triplet (V, E, p) where V is the set of nodes, E is the set of edges and $p : E \to (0, 1]$. The main challenge posed by this formulation is that problems that are relatively easy to solve in exact graphs become very difficult (or even intractable) in probabilistic graphs. In this paper, we perform an overview of the algorithmic techniques proposed in the literature for uncertain graph analysis. In particular, we center our focus on the following graph mining tasks: clustering, maximal cliques, k-nearest neighbors and core decomposition. We conclude the paper with a short discussion related to distributed mining of uncertain graphs which is expected to achieve significant performance improvements.

Keywords: Graph mining · Network analysis · Uncertain graphs

1 Introduction

Graph mining is an important research area with a plethora of practical applications [1,11]. The main reason for this is the fact that graphs are ubiquitous and, therefore, their efficient management and mining is necessary to guarantee fast and meaningful knowledge discovery.

In its simplest form, a graph $G(V, E)$, is composed of a set of nodes V, representing the entities (objects), and a set of edges E, representing the relationships among the entities. An edge $e_{u,v} = (u, v) \in E$ connects a pair of nodes u, v, denoting that these nodes are directly related in a meaningful manner.

© Springer International Publishing AG 2017
T. Sellis and K. Oikonomou (Eds.): ALGOCLOUD 2016, LNCS 10230, pp. 87–116, 2017.
DOI: 10.1007/978-3-319-57045-7_6

For example, if nodes represent authors, then an edge between two authors may denote that they have collaborated in at least one paper. As another example, in a social network application, an edge may denote that two users are connected by a friendship relationship.

A special category of graphs, include graphs that introduce *uncertainty* with respect to the existence of nodes and edges. For example, an edge e between nodes u and v may exist or not. The existence of an edge depends on several factors depending on the particular application under consideration. For example, in a social network where the edge corresponds to a message exchange between two users, the message will be sent with some probability (i.e., it is not sure that user u will send a message to user v). As another example, consider a protein-protein interaction network, where each node corresponds to a protein and each edge denotes that two proteins are combined. In this case, we may realize that proteins u and v interact in 70% of the cases, which means that the edge $e_{u,v}$ will be present in the graph with a probability of 0.7.

Let $\mathcal{G} = (V, E, p)$ be an uncertain (a.k.a probabilistic) graph, where $p : E \rightarrow (0, 1]$ is a function that assigns probabilities to the edges of the graph[1]. A widely used approach to analyze uncertain graphs is the one of *possible worlds*, where each possible world constitutes a deterministic realization of $\mathcal{G}$. According to this model, an uncertain graph $\mathcal{G}$ is interpreted as a set $\{G = (V, E_G)\}_{E_G \subseteq E}$ of $2^{|E|}$ possible deterministic graphs [41, 42]. Let $G \sqsubseteq \mathcal{G}$ indicate that G is a possible world of $\mathcal{G}$. Then, the probability that $G = (V, E_G)$ is observed as a possible world of $\mathcal{G}$ is given by the following formula:

$$\Pr(G|\mathcal{G}) = \prod_{e \in E_G} p(e) \prod_{e \in E \setminus E_G} (1 - p(e)) \qquad (1)$$

For instance, consider the probabilistic graph $\mathcal{G}$ shown in Fig. 1(a). Two possible instances of $\mathcal{G}$ are given in Fig. 1(b) and (c). Edges with high probability values are expected to show up more frequently in instances of $\mathcal{G}$. Consequently, triangles formed by high probability edges are more likely to be present in a random instance of $\mathcal{G}$. For example, it is not a surprise that the triangle formed by nodes v_4, v_5 and v_6 that has an existential probability of $0.9 \cdot 0.9 \cdot 0.9 = 0.729$ appears in both G_1 and G_2. In contrast, the triangle (v_1, v_4, v_5) has an existential probability of $0.5 \cdot 0.2 \cdot 0.9 = 0.09$ and its presence in a random instance of $\mathcal{G}$ is not very likely.

The annotation of edges with existential probabilities has significant impact on the algorithmic efficiency for particular problems. The uncertain existence of edges poses severe difficulties in solving problems whose counterparts in conventional (i.e., certain or deterministic) graphs can be easily addressed using polynomial-time algorithms. For example, assume that we are interested in determining the probability that nodes v and u are reachable from each other within a distance threshold. In a conventional (i.e., certain) undirected graph, the nodes

---

[1] Although existential probabilities can be assigned to the vertices of the graph as well, in this paper we focus on edge probabilities only.

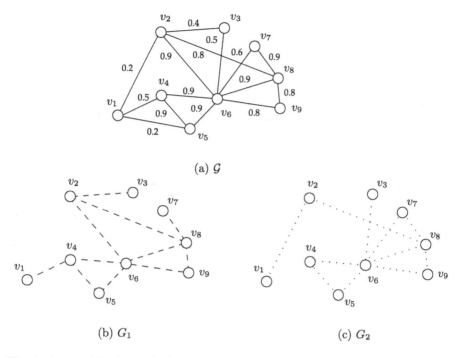

Fig. 1. A probabilistic graph $\mathcal{G}$ and two possible instances G_1 and G_2. The numbers near the edges denote existential probabilities.

will be either reachable if they belong to the same connected component or otherwise unreachable. Moreover, computing shortest path distances is considered a common graph operation that can be solved in polynomial time using Dijkstra's shortest path algorithm. On the other hand, computing the probability that the nodes are reachable in a probabilistic graph requires significant effort, since the problem is known to be #P-complete. Similar difficulties appear in other problems such as finding nearest neighbors or computing shortest paths in a probabilistic setting.

Based on the possible worlds semantics, several graph analytic tasks have been considered recently under the uncertain graph model. Note that, a simplistic technique to work with uncertain graphs is to assume that edge probabilities are simple weights. However, this approach does not produce meaningful results in many cases, since edge probabilities have different semantics than simple edge weights. Moreover, additional difficulties may be posed when edges contain existential probabilities and weights. Therefore, the recent years specialized algorithmic techniques have been proposed. The problem of *k-nearest-neighbors* have been addressed in [46], where shortest paths are computed based on a probabilistic approach. Another important graph mining task is *clustering* where graph nodes must be assigned to clusters based on connectivity. This problem is studied in [25,28]. Concerning the problem of mining *dense components* in uncertain graphs, [62] finds the densest induced subgraph in terms of the maximum expected density, that is the expected density value of an exact graph chosen at random. Another important concept is *reachability analysis* in uncertain graphs, where we are interested in determining if nodes are reachable given specific constraints [27]. Bonchi et al. [7] proposed an extension of the *core decomposition* for uncertain graphs. Very recently, a *triangle-based* extension of the core decomposition, namely the *truss decomposition*, was introduced for the uncertain graph model [24,65]. Other related contributions include algorithms for *subgraph similarity search* [60], *centrality computation* [45], the discovery of *frequent subgraph patterns* in uncertain graph databases [40].

In this paper, we focus on a subset of the aforementioned algorithmic techniques. In particular, we address the following graph mining problems: (1) clustering the nodes of an uncertain graph, (2) computing k-nearest-neighbors, (3) finding maximal cliques and (4) computing the core decomposition. In addition, we provide a short discussion related to how uncertain graph mining could benefit from distributed computation in clusters and what are the basic challenges that must be addressed in such a setting.

2   Clustering Uncertain Graphs

2.1   Introduction

Clustering has a plethora of applications in many diverse fields and it is considered as one of the most important data mining tasks in general. It is defined as the problem of grouping data objects into clusters, such that objects with similar characteristics are assigned to the same cluster and objects that have dissimilar characteristics are assigned to different clusters. In many cases, the clustering problem in graphs is highly related to *community detection* [15] which has been studied thoroughly in network analysis. The problem becomes extremely challenging when probabilities are assigned to the edges of the graph.

The problem of clustering probabilistic graphs has been recently studied by [28]. Simple techniques and methods that are used for partitioning graphs into clusters cannot be applied to probabilistic graphs, due to their nature. The challenges that arise are related to the applicability of the standard algorithms and the time complexity associated with some already developed algorithms. Many studies for uncertain data management and graph mining have been conducted in the computer science community [8,38]. These studies approach the probabilistic graph clustering problem as a deterministic one, by either considering the edge probabilities as weights or by leaving out probabilities smaller than a specific threshold. For the first approach the main problem is that it cannot solve the clustering problem for weighted probabilistic graphs, because once the probability is considered as weight, the actual weights cannot be encoded meaningfully onto the edges. For example, finding a mixed weight by multiplying the actual weight with the probability of the edge, gives a result with no possible real interpretation. For the second approach, the problem is that the threshold value cannot be computed in a principled, reliable way.

The probabilistic clustering problem has many important applications, with the most well-known being the discovery of complexes in protein-protein interaction networks and discovering communities in affiliation and social networks. The aforementioned problems contain uncertainty, so they are best represented using probabilistic graphs. In these cases, the nodes represent a protein or a user, and the edges represent the probability of an interaction between proteins or users.

The possible world semantics is being used to treat this type of graphs, meaning that every probabilistic graph $\mathcal{G}$ is treated as a generative model for deterministic graphs. This means that every possible instance of $\mathcal{G}$ represents a deterministic graph, or other words a possible world of $\mathcal{G}$. Figure 2 illustrates an example of a simple probabilistic graph $\mathcal{G}$ which consists of five nodes and eight edges.

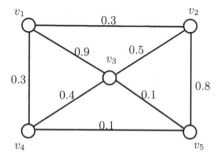

Fig. 2. A simple probabilistic graph $\mathcal{G}$ with 5 nodes and 8 edges

We call $D(G, C)$ the objective function that measures the cost of clustering the partitioning C of a deterministic graph $G = (V, E)$. Due to the possible world semantics, the expected value of this function D for clustering a probabilistic graph is the expected value of $D(\mathcal{G}, C)$ for all possible worlds of $\mathcal{G}$. By this definition, it is easily concluded that the computational cost of this function can be high, due to the number of possible worlds that are generated exponentially. There are $2^{|E|}$ possible graphs because each edge either will be part of the graph or not. Another problem with this approach is that there may be possible worlds with disconnected cliques, and therefore, it is very difficult to find a well-established clustering objective function, like the maximum cluster diameter or others [30], because the value will be infinite.

2.2   Algorithms

New definitions of the clustering problem in probabilistic graphs were proposed in [28]. The use of the edit distance as the optimization function is utilized for this definition. Considering a cluster graph C, i.e., a graph where there can be parts that are not connected with each other, as a clustering of a deterministic graph G, the edit distance between C and G is the number of edges that need to be added and removed from G, to get C. To adapt this approach for finding the cluster graph C for probabilistic graphs, a generalization of the CLUSTEREDIT problem by Shamir [52] was defined as the PCLUSTEREDIT problem. This problem can be defined as finding the clustering of $\mathcal{G}$, namely the cluster graph C, with the minimum expected edit distance from $\mathcal{G}$.

The advantages of this framework include its polynomial computational cost, the measurable output of the objective function and the independence of the clustering from factors other than the initial graph. This means that we do not need to evaluate each possible world of $\mathcal{G}$, so the output of the function will never be infinite because it is dependent on the node pairs within the graph. Furthermore, the variance of the output to any clustering is independent of the specific clustering but correlates only with the initial graph. Finally, the function does not require the specification of any free parameters, which means the number of clusters is part of the output.

Before getting to the algorithms, it is useful to explain the probabilistic graph model under consideration. The probabilistic graph $\mathcal{G}$ is defined as a tuple $\mathcal{G} = (V, P, W)$, where V is the set of nodes and $|V| = n$, P is the set consisting of every pair of nodes with a probability between 0 and 1; P_{uv} is the existential probability of the edge (u, v), and $|P| = m$. Finally, W is the sets of weights for all pairs of nodes. The algorithms presented below have been tested in undirected, unweighted probabilistic graphs and assume independence among edges, so $\mathcal{G} = (V, P)$. The complement of $\mathcal{G}$ is $\mathcal{G}'$ where $\mathcal{G}' = (V, 1 - P)$.

Given two deterministic graphs $G = (V, E_G)$ and $Q = (V, E_Q)$, the edit distance between G and Q is defined as the number of edges that need to be added to or deleted from G in order to be transformed into Q:

$$D(G, Q) = |E_G \setminus E_Q| + |E_Q \setminus E_G|. \tag{2}$$

By using the binary adjacency matrices $\mathbf{G}$ and $\mathbf{Q}$ of G and Q respectively we have:

$$(G, Q) = \sum_{u=1, v<u}^{n} \mathbf{G}(u, v) - \mathbf{Q}(u, v)|. \tag{3}$$

This definition is extended for a probabilistic graph $\mathcal{G}$ and a deterministic graph Q as the expected edit distance between every cluster $G \in \mathcal{G}$ and Q:

$$D(G, Q) = E_{G \subseteq \mathcal{G}}[D(G, Q)] = \sum_{G \subseteq \mathcal{G}} \Pr[G] D(G, Q). \tag{4}$$

Although this requires the calculation of all $2^{|E|}$ possible worlds, it is proven that the expected edit distance can be measured in polynomial time by using the following equation:

$$E_{G \subseteq \mathcal{G}}\left[\sum_{u<v} X_{uv}\right] = \sum_{u<v}(E_{G \subseteq \mathcal{G}} X_{uv}) = \sum_{\{u,v\} \in E_Q}(1 - P_{uv}) + \sum_{\{u,v\} \notin E_Q} P_{uv} \tag{5}$$

where X_{uv} denotes the random variable $|\mathbf{G}(u, v) - \mathbf{Q}(u, v)|$.

A *cluster graph* $C = (V, E_C)$ is a deterministic graph with the following properties: (1) C defines a partition of the nodes in V into k parts, $V = (V_1, \ldots, V_k)$ such that $V_i \cup V_j = \emptyset$. (2) For every $i \in (1, \ldots, k)$ and for every pair of nodes $v \in V_i$ and $v' \in V_i$ we have that $(v, v') \in E_C$. (3) For every $i, j \in (1, \ldots, k)$ with $i \neq j$ and every pair of nodes v, v' such that $v \in V_i$ and $v' \in V_j$, $(v, v') \notin E_C$. The PCLUSTEREDIT clustering problem is defined as follows: Given a probabilistic graph $\mathcal{G} = (V, P)$ find the cluster graph $C = (V, E_C)$ such that $D(\mathcal{G}, C)$ is minimized.

In the *correlation clustering problem* [5], there is a positive (+) or negative (−) relation between any two pair of objects. The notation E^+ is used for the set of pairs that are positively related and E^- for the corresponding negative. The goal is to find a partition that covers all V and minimizes the number of disagreement pairs, i.e., + pairs that are in different clusters and − pairs that are in the same clusters. When weights exist, the cost of clustering is the sum of W_{uv}^+ over all $\{u, v\}$ that are in different clusters plus the sum of W_{uv}^- for all the pairs that are in the same clusters. If all W are between 0 and 1 we have that $W_{uv}^+ + W_{uv}^- = 1$ thus, they satisfy the probability constraint and the objective function is formulated as:

$$CC(\mathcal{P}) = \sum_{(u,v), P(u)=P(v)} W_{uv}^- + \sum_{(u,v), P(u)\neq P(v)} (1 - W_{uv}^-) \qquad (6)$$

which is very similar to the objective function of the PCLUSTEREDIT problem. The only difference is that we need to replace $(1 - P_{uv})$ with W^+ and P with $(1 - W^-)$. The similarity of these problems showed that the approximation algorithm for the aforementioned problem can be utilized for the PCLUSTEREDIT problem as well.

Algorithm pKwikCluster. The first algorithm that we present is the PKWIK-CLUSTER algorithm that is originated from the KWIKCLUSTER algorithm [2] for the weighted correlation clustering problem. The algorithm starts by picking a random node u from the set of V. Then, it places u in the same cluster with all the nodes that are connected to it with probability higher than 0, 5. If u has not an edge with such a probability, it defines a singleton cluster, i.e., a cluster with only one node and no edges. Then the algorithm removes the nodes that belong to the newly formed cluster and repeats the process for the remaining nodes of the graph. The process is depicted in Algorithm 1.

As KWIKCLUSTER is a randomized 5-approximation algorithm for the correlation clustering problem we have that PKWIKCLUSTER algorithm is a randomized algorithm for the PCLUSTEREDIT problem. The algorithm has a time complexity of $O(n)$ since it only depends on the number of nodes and thus, it is linear and easily scalable. Figure 3 shows the result of one iteration of the algorithm for the example graph shown in Fig. 2.

Algorithm Furthest. The FURTHEST algorithm is a top-down algorithm that uses a center-based logic for the graph clustering. First, the algorithm assigns

Algorithm 1. PKWIKCLUSTER algorithm for probabilistic graph clustering.

repeat
    Choose $u \in V$ **randomly**
    $C(u) \leftarrow u$
    for all $v \in V$ such that $p(u,v) \geq 0.5$ **do**
        $C(u) \leftarrow C(u) \cup v$
    end for
    $V \leftarrow V - C(u)$
until $V = \emptyset$

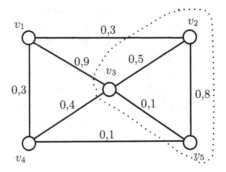

Fig. 3. The probabilistic graph after the first iteration of PKWIKCLUSTER. Node v_2 was picked randomly to be the center of the new cluster, and nodes v_5 and v_8, which they have an edge with probability equal or over 0.5 with v_2, were assigned to it.

every node into a single cluster. Then, it determines the pair of nodes that have the smallest probability of having an edge between them and marks these nodes as the centers of two new clusters. The remaining nodes are assigned to the cluster that are connected with the highest probability. This is an iterative process and at the end of each iteration i, a new cluster graph C_i is created with cost $D(\mathcal{G}, C_i)$. If $D(\mathcal{G}, C_i) < D(\mathcal{G}, C_i - 1)$ then the algorithm proceeds to the next iteration. Otherwise, it terminates and $C_i - 1$ is considered as the algorithm's output. The outline of this approach is depicted in Algorithm 2.

In each iteration, the algorithm computes the distance of each node to the existing cluster centers. If the output C consists of k clusters, then the complexity of the algorithm is $O(mk^2)$. However, the complexity can drop to $O(mk)$ if some distance caching is achieved in order not to recompute distances from previous pivots Fig. 4 shows the result of the execution of one iteration of the algorithm on the graph shown in Fig. 2.

Algorithm Agglomerative. The AGGLOMERATIVE algorithm is a bottom-up procedure for the PCLUSTEREDIT problem. First we must define a new term, called the *average edge probability*. The average edge probability between two clusters V_1 and V_2, is calculated by the formula:

$$\frac{1}{|Vi||Vj|} \sum_{u \in V_i, v \in V_j} P_{uv}. \tag{7}$$

Algorithm 2. FURTHEST algorithm for probabilistic graph clustering.

repeat
    $\mathcal{C} \leftarrow \emptyset$, $\mathcal{C} \subset V$ is the set of nodes acting as cluster centers
    for all $u \in V$ **do**
        $C(u) \leftarrow u$
    end for
    First iteration:
    $\mathcal{C} \leftarrow \mathcal{C} \cup \{c_1, c_2\}$, such that $c_1, c_2 \in V - \mathcal{C}$ and $p(c_1, c_2)$ is minimum
    i-**th iteration:**
    $\mathcal{C} \leftarrow \mathcal{C} \cup c_i$, such that $c_i \in V - \mathcal{C}$ and the probability between c_i and members
  of $\mathcal{C}$ is minimum
    for all $u \in V - \mathcal{C}$ **do**
        Assign u to the cluster with which it is more probable to share an edge
        $V \leftarrow V - \{u\}$
    end for
until $V \leftarrow \emptyset$

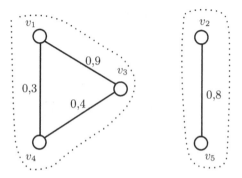

Fig. 4. The probabilistic graph after the first iteration of Furthest. Nodes v_4 and v_5 are picked as the new centers because they have the smallest probability between them. The remaining nodes are assigned to the center which they are closer to. The edges between the two clusters are removed.

The outline of this technique is depicted in Algorithm 3.

First, the algorithm turns every node into a singleton cluster. The algorithm is iterative. In each iteration i, it is checked If the largest average edge probability is more than $0, 5$. If it is, it places the two nodes with the largest average edge probability between them in C_{i-1} into the same cluster, making the cluster graph C_i. If not, it stops and outputs the previous clustering C_{i-1}. The complexity of the algorithm using a naive method is $O(km^2)$, where k is the number of clusters in the final clustering. By using a data structure like a heap for retrieving the closest pair of clusters and placing every edge in it, the complexity reduces to $O(km \log m)$. Figure 5 shows the result of one iteration of the algorithm on the simple graph illustrated in Fig. 2.

Based on the results provided AGGLOMERATIVE and FURTHEST do not scale well for large graphs. Also, the pKWIKCLUSTER, which is linear with respect

Algorithm 3. Agglomerative algorithm for probabilistic graph clustering.

repeat
    for all $u \in V$ **do**
        u forms a singleton cluster
    end for
    for all pairs of clusters **do**
        Find pair with maximum average edge probability p_{ae}
        if $p_{ae} \geq 0.5$ **then**
            Merge pair of clusters into one and continue
        else
            Stop and display current clustering
        end if
    end for
until $p_{ae} < 0.5$

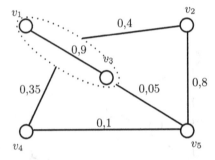

Fig. 5. The probabilistic graph after the first iteration of the Agglomerative algorithm. Nodes v_1 and v_3 are first selected to form a cluster as they have the largest edge probability between them. After that, the new probability is calculated for the newly formed cluster and nodes v_2 v_4 and v_5 using the average edge probability definition.

to the number of edges, exhibits the problem of including only edges associated with a probability of existing that is more than 0.5, which is sometimes not desirable or practical. Also, the output clustering of PKWIKCLUSTER does not contain any path longer than two edges, which is usually not desirable. As a result of this limitations, it would be useful to conduct more studies to ensure better clustering quality using scalable algorithmic techniques.

Algorithm Balls. The *Balls* algorithm was inspired by another algorithm introduced in [2], developed for the correlation clustering problem. The input of the algorithm is the matrix of distances between pairs of edges $\mathcal{X}_{uv}$. The algorithm uses a parameter a that is set to a constant for a constant approximation ratio, but it can be changed accordingly depending on the problem at hand.

    The BALLS algorithm tries to find a set of nodes that are close to each other and far from other nodes. If a set with this characteristic is discovered, it removes it from the graph as a cluster and continues with the rest of the graph. To discover such a set of nodes is not easy, because every subset of nodes of the probabilistic graph $\mathcal{G}$ must be considered. To solve this problem, [19] used the triangle inequality method for distance X_{uv}. The algorithm tries to find clusters that are close (within a ball), to a node u, using the guarantee of triangle inequality, which concludes that if two nodes are close to u, then they are close to each other too. This algorithm tends to output ball-shaped clusters.

The steps of the algorithm are the following: first the nodes are placed in increasing order of the total sum of possibilities incident to each node. At every iteration, the algorithm selects the first unclustered node u of the ordered set. Then, it finds the set of nodes B that have probability a of 0.5 or more to have an edge with u. Then, the average distance $d(n, B)$ of the nodes in B from node u is calculated. If $d(n, B) < a$, then we get a cluster with all the nodes from B and node u. Else, u is a singleton cluster. The space complexity of the algorithm is $O(mn^2)$ for generating the table and its time complexity is $O(m^2)$.

To evaluate these algorithms, experiments were conducted that included the core PPI network. This dataset that we will refer to as COREPPI was provided by [31]. It contains 2708 nodes that represent proteins and 7123 edges that represent protein interactions. The edge probabilities show how likely it is that the interaction actually happens between two proteins. About 20% of the edges have probability over 0.98, while no edge has probability less than 0.27 in the graph. The remaining edge probabilities are uniformly distributed in the remaining range $[0.27, 0.98]$. Finally, the dataset is characterized by power-law degree distribution, short paths and high clustering coefficient.

The experiment that was conducted by [28], aimed to compare the performance of PKWIKCLUSTER, AGGLOMERATIVE, FURTHEST and BALLS algorithms regarding their running time and quality of output. The latter is measured in accordance with the objective function, which for this experiment was the edit distance of the output cluster graphs from the input probabilistic graphs, as we have explained before.

We present in Table 1 the results of the algorithms regarding their running time and the expected edit distance. The best value of the objective function for PKWIKCLUSTER is reported, after running it 100 times, since it is a randomized algorithm. Regarding the objective function, aka the expected edit distance we get the best result from AGGLOMERATIVE with 3420, followed by PKWIK-CLUSTER with 4194. Regarding running time, PKWIKCLUSTER is the fastest as expected with 0.005 s running time, since it is linear to the number of the input edges. It must be noted that although the dataset is small, the FURTHEST algorithm takes more than a minute to finish, showing its scalability problem. Actually, all algorithms except PKWIKCLUSTER cannot scale to more than a few thousand nodes. The REFERENCE clustering has time complexity at least quadratic to the number of nodes because it uses matrix multiplications and the reported results are given from [31].

The known ground truth that was used to validate the results of the algorithms is the MIPS database [35]. MIPS complexes define relationships among proteins of the same complex and this knowledge is utilized only for the **Gcore**

Table 1. Summary of results.

Algorithm	Complexity	Distance	Runtime	# Clusters	TP	FP	FN
REFERENCE	N/A	12230	N/A	547	1791	11635	3589
BALLS	$O(m^2)$	4960	8	757	1120	1734	4260
PKWIKCLUSTER	$O(n)$	4194	0.005	757	1120	1734	4260
AGGLOMERATIVE	$O(kmlogm)$	3420	10	542	946	1357	4434
FURTHEST	$O(mk^2)$	4612	150	619	894	2322	4486

graph data and the result is 5380 pairs of proteins. The key difference is that this includes proteins that belong to more than one cluster, but the algorithms output cluster partitions, meaning that every protein can be only in one cluster.

Table 1 shows the number of non-singleton clusters that occurred as a result of each algorithm, and also the True Positive edges, i.e., edges that appear at the clustering and also in MIPS ground truth, the False Positive edges, i.e., edges that appear at the clustering but they do not exist in MIPS ground truth, and finally the False Negatives, i.e., edges that do not appear at the clustering, but they exist in MIPS ground truth. It can be observed that AGGLOMERATIVE and FURTHEST results for non-singleton clusters are the closest to REFERENCE. Also it can be noticed that every algorithm produces quite different results with different trade-offs. For example, PKWIKCLUSTER output includes only 838 TP edges, while REFERENCE contains 1791. However, PKWIKCLUSTER has better results than REFERENCE regarding the FPs.

2.3 Reliable Clustering

The possible worlds model also comprised the basis of the algorithm proposed in [33] for *reliable clustering* of probabilistic graphs. Reliability in terms of clustering involves the connectivity of the clusters over different possible worlds of the probabilistic graph, and two metrics were proposed for the measurement of the reliability of a clustering. A probabilistic alteration of the classic k-means was proposed based on the two new metrics, which is described below.

In deterministic (i.e., certain) graphs, the connectivity of a clustering is typically measured by the sum of the weights of the edges between each pair of clusters. Using this metric for probabilistic graph clustering where the probabilities of the edges are perceived as weights can be problematic, as it does not take into account the possible world semantics. Another problem raised is that the connectivity of a cluster could be influenced by nodes belonging to different clusters, instead of being solely dependent on the nodes it contains. Thus, the criteria for probabilistic graph clustering should capture both local and global relationships between the nodes of the graph.

The *standard uncertain graph reliability* proposed in [10] poses such a criterion for reliable probabilistic graph clustering, but it can be very computationally expensive, which lead to the proposal of the *generalized reliability criterion*.

In both cases, the reliability $R(C)$ of a set of vertices C, which comprises a cluster subgraph of $\mathcal{G}$, is defined as follows:

$$R(C) = \sum_{G_i \sqsubseteq \mathcal{G}} \Pr(G_i)\mathcal{I}(C, G_i) \qquad (8)$$

where $\mathcal{I}(C, G_i)$ is equal to 1 if the cluster C is contained in a connected component in G_i, and 0 otherwise, and G_i is a possible world derived from $\mathcal{G}$.

In this sense, the reliability metric extends the concept of connectivity from deterministic graphs to probabilistic graphs and measures the probability of a probabilistic clustering to maintain its vertices connected in some possible world.

The two basic intuitions behind the generalized reliability criterion are described as *purity* and *size balance*. Given a possible world G of $\mathcal{G}$, purity imposes the constraint that in each connected component of G the number of different clusters should be minimal and one of the clusters should dominate the component. In other words, each connected component in this possible world should be pure, in the sense that the nodes contained in it should exhibit similar characteristics. The purity metric can then be formulated using the cluster label entropy:

$$\mathcal{F}_p = \sum_{G_i \sqsubseteq \mathcal{G}} \Pr(G_i) \sum_{j=1}^{L_i} |CC^{i,j}| H(\bigcup_k CC_k^{i,j}) \qquad (9)$$

where $CC^{i,j}$ denote the L_i connected components in G_i, $CC_k^{i,j}$ denote the nodes belonging to the k-th cluster and $H(\bigcup_k CC_k^{i,j})$ is the entropy of cluster labels for the j-th connected component of G_i, defined as:

$$H(\bigcup_k CC_k^{i,j}) = -\sum_{k=1}^{K} \frac{|CC_k^{i,j}|}{|CC^{i,j}|} \log \frac{|CC_k^{i,j}|}{|CC^{i,j}|}. \qquad (10)$$

The above criterion biases the clustering algorithm towards the selection of a single clustering containing all or most of the nodes. This is the reason behind the imposition of the second metric, which is described as size balance. The size balance metric is formulated as follows:

$$\mathcal{F}_e = \sum_{G_i \sqsubseteq \mathcal{G}} \Pr(G_i)|V|H(\bigcup_k C_k) = |V|H(\bigcup_k C_k) \qquad (11)$$

where $H(\bigcup_k C_k)$ is the entropy of cluster size, defined as:

$$H(\bigcup_k C_k) = -\sum_{k=1}^{K} \frac{|C_k|}{|V|} \log \frac{|C_k|}{|V|} \qquad (12)$$

Given the above definitions, the generalized reliability criterion guides the clustering algorithm towards the minimization of $\mathcal{F} = \mathcal{F}_p - \mathcal{F}_e$. This objective is equivalent to the minimization of the following equation:

$$\mathcal{F} = -\sum_{G_i \sqsubseteq \mathcal{G}} \Pr(G_i) \sum_{j=1}^{L_i} \sum_{k=1}^{K} |C_k^{i,j}| \log(\frac{|C_k^{i,j}|}{|C_k|}) \qquad (13)$$

A Monte-Carlo sampling technique is then used to create N possible worlds of $\mathcal{G}$, denoted by $G_i, i = 1, \ldots, N$, in order to estimate the reliability criterion. In this fashion, the generalized reliability criterion can be reformulated as follows:

$$\mathcal{F}_s = \frac{1}{N} \sum_{i=1}^{N} \sum_{k=1}^{K} |C_k| H(\bigcup_j CC_k^{i,j}) \tag{14}$$

where $\mathcal{F}_s$ is an unbiased estimator of $\mathcal{F}$.

Moreover, an auxiliary cluster table $\mathcal{T}_i$ is defined for each possible world G_i, in which each row represents a cluster and each column a connected component. Then, $\mathcal{T}_i(k, j)$ contains the $CC_k^{i,j}$ set of vertices, as defined above. The definition of $\mathcal{T}_i$ allows for the utilization of various coding algorithms for the purpose of encoding each table corresponding to each possible world.

With the rows of the auxiliary table representing each of the k clusters and using a fixed coding, each vertex v can be assigned to the row-cluster which reduces its coding cost. The algorithm begins by assigning each sample randomly to one of the clusters. Then, the coding for each cluster distribution is computed, using the Huffman coding or any other type of coding algorithm. After this computation, each vertex v is assigned to the k^*-th cluster, which minimizes its coding cost over all possible worlds. It has been proven that through this iterative process $\mathcal{F}$ converges to a (local) minimum. The outline of the process is depicted in Algorithm 4.

Algorithm 4. Coded k-means algorithm for probabilistic graph clustering.

Require: $CC^{i,j}$: connected components, K: number of clusters
   $C_k \leftarrow \emptyset, \forall k$
   for all $v \in V$ **do**
      $C_k \leftarrow C_k \cup v$, k is chosen randomly from $\{1, \ldots, K\}$
   end for
   while $\mathcal{F}$ hasn't converged **do**
      for all i **do**
         for all k **do**
            $c(CC_k^{i,j}) \leftarrow HuffmanCoding(CC_k^{i,j})$
         end for
      end for
      $C_k = \emptyset$
      for all $v \in V$ **do**
         for all i **do**
            for all k **do** $CodeLength_k(v) += |c(CC_k^{i,f(v,i)})$
            end for
         end for
      end for
      $C_{k^*} \leftarrow C_{k^*} \cup \{v\}$, where k^* minimizes the coding cost $CodeLength_k(v)$
   end while
   return $C_k, k = 1, \ldots, K$

Experiments conducted using the DBLP dataset [46] and the PPI dataset [64] showed that the proposed k-means alteration improved performance in terms of the average coding length per vertex, the average vertex pairwise reliability and the average cluster reliability.

3   Clique Discovery in Uncertain Graphs

3.1   Introduction

One fundamental problem in graph mining is discovering vertices that are densely connected. A *clique* is a set C of vertices, where every vertice is connected by an edge, i.e. for every two vertices u and $v \in C$, there is an edge that connects the two vertices. A *maximal clique* is a clique that is not contained in any other clique. Cliques or maximal cliques can often be considered as cores of structures in graphs.

Discovering cliques in graphs is a fundamental task with many applications such as community detection in social or biological networks [39], genetics study under different circumstances [47] and genome mapping data integration [22].

The problem of finding top-k maximal cliques in a probabilistic graph has been studied in [63]. An application of this problem is found in protein-protein interaction networks, where studies has shown that cliques usually represent cores of protein complexes [4]. Because of the existence of probabilities on the edges of graphs, a set of vertices may not form a maximal clique in all possible worlds. The term *maximal clique probability* is used to describe the possibility that a set of vertices is a maximal clique in all possible worlds of the uncertain graph. A collection of k sets with the largest maximal-clique probabilities is defined as the top-k maximal cliques. Due to the fact that cliques which contain a small number of vertices do not give us useful information, we choose to discover cliques that contain at least s vertices.

The probability of a set of vertices C to be a maximal clique for all the possible worlds of an uncertain graph G, i.e. the *maximal-clique probability* of C is given by $\sum_{G' \in \Omega} \Pr(G \Rightarrow |G'|)$ where Ω is the set of graphs for all possible worlds and G' is a world in which C is a maximal clique.

The problem of finding the top-k maximal cliques is NP-hard because it contains the maximal cliques enumeration problem [36], where $k = \infty$, $s=1$ and G is an exact graph. The problem is defined for an uncertain graph $\mathcal{G}$ and two positive integers k and s. The final output of the problem is a collection F of k sets of vertices of $\mathcal{G}$ with the following qualities: (1) the size of each set in F is equal to or more than s and (2) for any set of cliques $C \in F$, and any other set $C' \notin F$, the maximal-clique probability of C is not less than that of C'.

3.2   Algorithmic Techniques

It has been proven in [63] that the maximal-clique probability of a set of vertices can be found in polynomial time. Considering this, we present the BRANCHANDBOUND algorithm that was proposed to find top-k maximal cliques in an uncertain graph.

Given a probabilistic graph $\mathcal{G}$ and G the deterministic graph that occurs if we remove the edge probabilities from $\mathcal{G}$, we consider $\prec$ to be the ascending order of indices of the vertices in $\mathcal{G}$. All of the cliques in G can be organized into a search tree, where the root represents a clique with no vertices, each node contains a unique clique in G and the parent of each non-root node represents another clique C' which has the following properties: (1) $C' \subset C$, (2) $|C| = |C'| + 1$ and (3) the only vertex $v \in C - C'$ satisfies $u \prec v$ for each vertex $v \in C$. So the problem of finding top-k maximal cliques in G can be approached as a tree searching problem, by trying to discover k nodes in the search tree that their maximal-clique probability is no less than that of any other node.

The BRANCHANDBOUND algorithm utilizes a min-heap H_{topk} of size k which stores the top-k nodes that have been found so far. The criterion that the nodes in H_{topk} are chosen is their maximal-clique probability, $\Pr(mcliq(C))$. The algorithm is initialized by storing a variable γ, which holds the maximal-clique probability of the root and its value is set to zero before the beginning of the algorithm. Also, a max-heap H_{ext} is used for the nodes that have not been examined yet. The nodes are stored in H_{ext} based on their clique probabilities $\Pr(cliq(C))$. At first, the singleton set u is stored into H_{ext}, for every v in G. Finally, the key of the singleton set v in the heap is $\Pr(cliq(v)) = P_V(v)$.

BRANCHANDBOUND in its basic form performs a simple best-first branch-and-bound search on the search tree. In each iteration, the algorithm takes the top node of H_{ext}, i.e., the node with the largest clique probability. For this node C, the algorithm applies 4 steps: **pruning**, a **computing**, an **expanding** and **updating**.

- During the **pruning step**, the algorithm checks if the H_{topk} is full and if $\Pr(cliq(C)) \leq \gamma$. If these do not hold true, then there might be top-k maximal cliques in the subtree that has as root C and the following steps of the algorithm must be performed. Otherwise, the aforementioned subtree can be pruned because the maximal-clique probability of each of its nodes is for sure less than $\Pr(cliq(C))$, so $\Pr(mcliq(C)) \leq \gamma$. This means that C' is not a top-k maximal clique and the algorithm can skip the following steps.
- In the **computing step**, the algorithm computes the maximal-clique probability of C. At first, it finds the set of vertices adjacent to all vertices in C, called $N(C)$. For each vertex $v \in N(C)$, the algorithm gets a new clique $C' = C \cup v$ and computes its $\Pr(cliq(C'))$. After that, it finds the maximal clique probability of C.
- In the **expanding step**, the algorithm checks if C' is a child of C in the search tree. If this holds true and if H_topk is full with $\Pr(cliq(C')) \leq \gamma$, then the subtree at C' can be pruned, otherwise the algorithm inserts C' into H_{ext} as a possible candidate for future searching.
- In the **updating step**, if $|C| > s$, then H_{topk} is updated in two possible ways: (1) If the heap is not full, C is inserted and (2) if it is full and $\Pr(mcliq(C)) > \gamma$, then the root of the heap is removed and C is inserted because the root of H_{topk} cannot be a top-k maximal clique.

Algorithm 5. BRANCHANDBOUND algorithm for top-k maximal clique search in probabilistic graphs.

---

repeat
    if $\mathcal{H}_{topk}$ is not full **and** $\Pr(cliq(C)) \leq \tau$ **then**
        for all $v \in V$ **do**
            $\mathcal{H}_{ext} \leftarrow \{v\}, key_{\mathcal{H}_{ext}}(v) \leftarrow \Pr(cliq(v))$
        end for
    $C \leftarrow$ root of $\mathcal{H}_{ext}$
    if $\mathcal{H}_{topk}$ is not full **and** $\Pr(cliq(C)) \leq \tau$ **then**
        $N(C) = \{v : \forall u \in C, (u, v) \in E\}$
        for all $v \in N(C)$ **do**
            $C' \leftarrow C \cup v$
            $\Pr(cliq(C')) \leftarrow \Pr(cliq(C)) * P_V(v) \prod_{u \in C} P_E((u, v)|(u, v)$
        end for
        Compute $\Pr(mcliq(C))$
        for all $v \in N(C)$ **do**
            if $u \prec v, \forall u \in C$ **then**
                C' is a child of C in the search tree
                if $\mathcal{H}_{topk}$ is full **and** $\Pr(cliq(C')) \leq \tau$ **then**
                    Prune subtree rooted at C'
                else
                    $\mathcal{H}_{ext} \leftarrow \mathcal{H}_{ext} \cup C'$
                end if
            end if
        end for
        if $|C| \geq s$ **then**
            if $\mathcal{H}_{topk}$ is not full **then**
                Insert C into $\mathcal{H}_{topk}$
            else if $\mathcal{H}_{topk}$ is full **and** $\Pr(mcliq(C)) > \tau$ **then**
                Remove root of $\mathcal{H}_{topk}$
                Insert C into $\mathcal{H}_{topk}$
            end if
        end if
    end if
    else
        Prune subtree rooted at C
    end if
until $\mathcal{H}_{ext}$ is empty **or** $\mathcal{H}_{topk}$ is full and $\Pr(cliq(C)) \leq \tau$

The algorithm terminates when either H_{ext} is empty or H_{topk} is full and $\Pr(cliq(C)) \leq \gamma$. The first condition is true when the whole non-pruned tree has been searched and the second condition is true when all subtrees below the nodes of H_{ext} can be pruned safely. The output of the algorithm consists of all the cliques in H_{topk}, which are the maximal cliques of $\mathcal{G}$. The outline of this technique is illustrated in Algorithm 5.

There is also an optimized version of the algorithm which uses techniques to improve efficiency. These techniques are: (1) size-based pruning, (2) look-ahead pruning and (3) anti-monotonicity-based Pruning. The optimized algorithm consists of two phases and can be further studied at [63].

Another approach for solving the maximal clique problem in uncertain graphs has been studied by [59]. The problem is tackled with the use of three metaheuristics, which are based on the popular tabu search (TS) [20], tabu's variation for discovering stable sets called STABULUS [16] and GRASP [13] algorithms. TS prevents getting stuck in local optima by using non-improving moves and the so-called *tabu list* and *tabu tenures* that forbid getting repetitive solutions. STABULUS differs from TS in the sense that it performs the local search on partially impracticable solutions. When the solution is found, the algorithm starts over to discover a better solution. The GRASP algorithm is a multi-start algorithm, in which the solution is found in a randomized greedy way in each iteration and then local search techniques are used to improve the current solution. The greedy part of the algorithm refers to the construction of the *restricted candidate list* which contains some of the best candidates. The algorithms developed using these techniques consider the CVAR (Conditional Value-at-Risk) [48] verification procedure, the robustness during local search moves and the bounds on the solution size.

3.3   Enumerating Maximal Cliques

Another problem, similar to finding the top-k maximal cliques in an uncertain graph, is called *maximal clique enumeration*. This problem has many applications in areas where data are more accurately represented by uncertain graphs like social networks [34, 44], email networks [43], protein-protein interaction networks [17] and bioinformatics [61].

The problem of enumerating maximal cliques in uncertain graphs has been studied in [36, 37]. It is used to find robust communities in graphs that contain probabilities. To solve this enumeration problem, an upper and lower bound for the largest number of maximal-cliques within a graph must be discovered.

First, the term a-maximal-clique is defined. An a-maximal clique is a maximal clique with probability at least a (where $0 \leq a \leq 1$)). If W is a set of vertices that form a clique with probability at least a, there is no other W' such that $W \in W'$, where W' a clique with probability at least a.

It is shown [36] that the maximum number of a-maximal-cliques in an uncertain graph with n vertices is $\binom{n}{\lfloor n/2 \rfloor}$. This means that there is an uncertain graph with $\binom{n}{\lfloor n/2 \rfloor}$ uncertain maximal cliques and no uncertain graph can have more than $\binom{n}{\lfloor n/2 \rfloor}$ a-maximal cliques.

The algorithm that is used for enumerating all a-maximal cliques is called MULE (Maximal Uncertain Clique Enumeration). The MULE algorithm is performing a depth-first-search (DFS) of the graph, utilizing optimization techniques for limiting search space and incremental computation of clique probabilities techniques for maximality check. The worst-case running cost of MULE on a graph with n vertices is $O(n2^n)$. However, this cost only appears when the graph is very dense, with much better results in typical graphs. It is proven that the algorithm does not perform an exhaustive search of the graph space and it can be optimized to find only large maximal cliques.

A simplistic approach for a-maximal cliques enumeration in a probabilistic graph $\mathcal{G}$ is to perform DFS with backtracking. At first, there is an empty set of vertices C that is an a-clique and the algorithm starts to add vertices to C, with the restriction that C must always be an a-clique, until no other vertices can be added. When that happens, we have an a-maximal-clique. Then, the algorithm backtracks to add any other possible vertices in C, until all possible search paths are explored.

The MULE algorithm improves this approach of DFS in several ways. First, the additional vertices that can be put in a current a-clique C have the property that are already connected with every vertex of C. So, it is efficient to try to discover this kind of vertices while the algorithm progresses, making the process quicker by not checking if a new vertex can be added to C. As a result, an incremental track of vertices is performed for the extension of C.

Another improvement of the MULE algorithm refers to the clique probability. If a vertex extends C into a clique, it does not mean that it extends C into an a-clique, too. The clique probability of C is decreased by a factor equal to the product of the edge probabilities between v and every vertex in C, when v is added in C. The algorithm calculates the factor by which the clique probability will fall, in $O(1)$ time, by incrementally maintaining the factor for each vertex v under consideration.

The last improvement that the MULE algorithm adds to the DFS approach is the reduction of maximality check. This is achieved by maintaining the set of vertices that can extend C, but will be explored in a different search path. This reduces the time cost from $\theta(n^2)$ to $\theta(n)$.

4   Nearest-Neighbor Search

4.1   Introduction

Another problem that has many applications regarding uncertain graphs is the *nearest-neighbor search* problem. Some of these applications include identifying protein neighbors in protein-protein interaction networks that is useful for possible co-complex membership predictions [31] and possible new interactions [51]. Another example refers to social networks, due to the uncertainty of their nature [51]. In social networks, it is useful to extract information from queries like how many people does a specific person influences the most with their on-line actions. In mobile ad-hoc networks, k-nearest neighbor queries can be used for connectivity applications [6] or for the *probabilistic-routing problem* [18]. The fundamental problem of computing distance functions and processing k-NN queries has long been studied for standard graphs, and it is also very important for the probabilistic graphs.

One way to compute the distance between two nodes v and u in a probabilistic graph is to consider the length of the MOST PROBABLE PATH (MPP). This distance is defined as the length of the path with the highest probability. Given

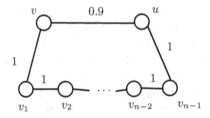

Fig. 6. Example of an arbitrarily long most probable path.

two nodes v and u in a probabilistic graph $\mathcal{G}$, we can consider two alternatives as indicators for the closeness of the nodes. The first is the length of the most probable path and the second is the probability that at least one path exist.

The MPP distance can be easily computed in a probabilistic graph by assuming that the graph is certain and by running the Dijkstra shortest-path algorithm. However, this approach presents several problems. The probability of such a path may be extremely small, and even if it is large, the probability that it is indeed the shortest path can be itself very small. Figure 6 illustrates an example where the lower path which has an arbitrary length n is the most probable path, whereas the direct path between v and u is a little less probable. These problems are solved in [46] by using statistics of the *shortest path distribution*.

4.2 Distance Measures

Considering a possible world G from $\mathcal{G}$, let $d_G(v,u)$ be the shortest path distance between v and u. The distribution $\mathbf{p}_{v,u}$ of the shortest path distance is defined as:

$$\mathbf{p}_{v,u}(d) = \sum_{G \mid d_G(v,u)=d} \Pr[G] \tag{15}$$

This is the sum of all the possible worlds in which the shortest path distance between v and u is equal to d. Because of the nature of the probabilistic graphs, sometimes there are worlds where v and u are belong to disconnected parts of the graph, thus making this distance equal to ∞. So $p_{v,u}(\infty)$ is defined as the total probability of all the worlds that v and u are disconnected.

Below we present different types of distances that will be used for different ways of k-NN pruning, studied in [63]. For these distances, the probabilistic graph $G = (V, E, P, W)$ and two any two nodes v and u will be used to define them.

Median Distance: The *Median Distance* $d_M(v,u)$ is the shortest path distance among all possible worlds. The median distance can be infinite for some pairs v and u and k-th order statistic are held as well. Formally:

$$d_M(v,u) = \arg\max_D \left\{ \sum_{d=0}^{D} \mathbf{p}_{v,u}(d) \leq \frac{1}{2} \right\} \tag{16}$$

The exact calculation of this distance is difficult to be executed, since it involves executing a point-to-point shortest-path algorithm in every possible world and taking the median. To solve this problem, we approximate the value of the median distance by sampling, using the Chernoff bound [46].

Majority Distance: The *Majority Distance* $d_J(v, u)$ is defined as the most probable shortest path distance. For weighted graphs, this distance has meaning if the weights come from a discrete domain. Formally:

$$d_J(v, u) = \arg \max_d \mathbf{p}_{v,u}(d) \tag{17}$$

Expected Reliable Distance: For this distance, only the possible worlds containing a path between v and u are considered. The *Expected Reliable Distance* $d_{ER}(v, u)$ is defined as the shortest path distance in all worlds in which there exists a path between v and u. This can be described formally as:

$$d_{ER}(v, u) = \sum_{d|d<\infty} \frac{\mathbf{p}_{v,u}(d)}{1 - \mathbf{p}_{v,u}(\infty)} \tag{18}$$

Computing this distance is considered a #P-hard problem, as a generalization of the reliability problem [57].

Another promising distance function has been defined, which is based on the concept of *random walks*. The difference between this approach and the shortest path approach is that it considers all possible paths, whilst the shortest path distance relies on one path only. Also, the random walk approach, as its name suggests, uses random instead of optimal choices. In standard graphs, random walks have been studied in [49]. The random walk distance function is inspired by the *Individual Page Rank* (IPR) concept [14]. For deterministic graphs, in a IPR walk that starts from a node v, it always teleports back to v, instead of teleporting to any node in the graph. For probabilistic graphs, IPR is defined considering a weighted probabilistic graph $\mathcal{G} = (V, E, W, P)$, where W represents the *proximity* between nodes in the graph.

The parameterization of the random walk is based on a source node s and a teleportation probability a. The walk starts at source node s and possible world G_0, which is sampled online according to P. At step i, we are at the node u_i and in the world G_i. At this step there are 2 choices: either follow an active edge with probability $1 - a$, or teleport to s with probability a. If there are no outgoing edges we stay at the same node. This process is called probabilistic random walk and the *random walk distance* is defined as the inverse of the stationary probability of u of the probabilistic random walk with starting node s.

4.3   Algorithms

Different k-NN algorithms have been developed based on the distance functions described above. First, let us present the definition of the k-NN problem. Given $\mathcal{G} = (V, E, P, W)$ a source node s, a probabilistic distance d_P, and a positive integer k corresponding to the number of neighbors, find the set of nodes $T_k(s) = (t_1, ..., t_k)$ for which the distance $d_P(s, t_i)$ is less or equal to the distance $d_P(s, t)$ for any other node $t \in V \setminus T_k(s)$. The problem lies in the computation of the set $T_k(s)$ without having to compute the distance $d_P(s, t)$ for all nodes $t \in V$, as this can be computationally expensive especially for large graphs.

Using the median distance as defined above, involves truncating the distribution $\mathbf{p}_{v,u}$ so as to contain distance values smaller than a given value D. The remaining values (for $d > D$ in the original distribution) are redistributed and concentrated exactly at $d = D$. Let the D-truncated distribution be $\mathbf{p}_{D,v,u}(d), d_{D,M}(v, u)$ be the median distance obtained from it and $d_M(v, u)$ be the actual median distance obtained from the original non-truncated distribution. It has been proven that if $d_{D,M}(v, u_1) < d_{D,M}(v, u_2)$ for any two nodes u_1, u_2, then $d_M(v, u_1) < d_M(v, u_2)$. In other words, the median distance obtained from the truncated distribution preserves distance relationships between the nodes of the graph. Thus, finding the set of nodes $T_k(v) = u_1, \ldots, u_k$ for which $d_{D,M}(v, u_i) \leq d_{D,M}(v, u)$ for $i = 1, \ldots, k$ and $v \in V - T_k(v)$ comprises a solution to the k-NN query for v.

The probability distribution is then approximated using sampling techniques. The algorithm based on the median distance begins by applying Dijkstra's algorithm with v as the starting node. Once a node gets traversed it is never been visited again, and to visit a new node, a sample of its outgoing nodes is generated. The algorithm terminates when it reaches a node whose distance is greater than D. Then, the sampled distribution of all the visited nodes is updated or instantiated. Performing the above steps r times yields r samples of the distribution. If the distribution of a node u reaches half of its mass, the node is added to the k-NN query result. Algorithm 6 presents the outline of the median-based variation described above.

Using the majority distance defined previously, a variation of the k-NN algorithm similar to the one described above is obtained. One difference lies in the way the k neighbors are collected. In the median distance version, once the truncated distribution of a node u reaches 50% of its mass, it is added to the solution. In the majority distance alternative, the condition ensuring that d_1 will be the majority distance is:

$$\tilde{\mathbf{p}}_{D,v,u}(d_1) \geq \frac{r - r_u}{r} \tag{19}$$

where d_1 is the current majority value in the sampled distribution $\tilde{\mathbf{p}}_{D,v,u}(d)$ and r_u is the number of times the node u has been visited during the r Dijkstra traversals.

Experiments conducted on the PPI and DBLP datasets showed that the above proposed algorithms showed improved performance in terms of true positives versus false positive rates, in comparison to random walk and reliability-based algorithms.

5   Core Decomposition

The *core decomposition* is a useful tool for performing a wide range of graph mining tasks. One of its advantages against other methods is that it can be

---

Algorithm 6. Median distance k-NN algorithm for nearest-neighbor queries in probabilistic graphs.

Require: $s \in V$: starting node, r: number of samples, k: number of neighbors, γ: distance increment
  $T_k \leftarrow \emptyset$
  $D \leftarrow 0$
  Run Dijkstra algorithm r times, from s
  while $|T_k| < j$ **do**
    $D \leftarrow D + \gamma$
    for $i \leftarrow 1 : r$ **do**
      Visit nodes in the i-th Dijkstra's execution until distance D is reached
      for all $t \in V$ visited by Dijkstra **do**
        update $\tilde{\mathbf{p}}_{D,s,t}$
      end for
    end for
    for all $t \notin T_k : \tilde{\mathbf{p}}_{D,s,t}$ exists **do**
      if $\tilde{\mathbf{p}}_{D,s,t} < D$ **then**
        $T_k \leftarrow T_k \cup \{t\}$
      end if
    end for
  end while
  return T_k

---

computed efficiently (linearly) in the size of the input graph. The core decomposition is related to the problem of discovering *dense subgraphs*, like *cliques*, *lambda-sets*, etc., most of which are NP-hard or they have high computational complexity and therefore, the ability to solve this problem in linear time is very appealing.

The k-core of a graph is defined as a maximal subgraph in which every vertex is connected to at least k other vertices within that subgraph [7]. The set of all these k-cores in a graph G forms the *core decomposition* of G [50]. While this problem is solved in linear time in deterministic graphs, it does not mean that it is solved this efficiently in probabilistic ones, due to their nature.

Core decomposition has been used to speed-up the computation of other problems with purpose of finding dense subgraphs in deterministic graphs. Some examples of this contribution are in the maximal-clique discovery problem [12], the *densest subgraph problem* [29] and the *densest at-least-k-subgraph* problem [3]. A core-decomposition solution developed for uncertain graphs would provide a natural extension of the aforementioned applications. Some applications of core decomposition in probabilistic graphs include *influence maximization* [58] and *task-driven-team-formation* [23].

In *influence maximization*, we have edges with probabilities that represent influence between nodes, and the goal is to find the nodes, e.g., users in a social network, that have the highest influence over a large number of users. To solve this problem, a greedy algorithm [21] has been proposed that requires a number of Monte Carlo simulations, which is computationally expensive. By using the core decomposition technique, this process can be executed more efficiently.

In *task-driven team formation*, the input is a collaboration graph $G = (V, E, r)$, where vertices represent individuals and edges represent topic(s) of past collaboration with the use of the probabilistic model r. Given a query $Q = (T, V)$, where T is a set of terms used to describe a new task and V is a set of individuals represented by vertices, the goal is to find the best set of vertices $A \subset Q$, to perform the task T. Due to the nature of the problem, we have a single probability for each edge pair $(v, u) \in E$, that (v, u) collaborate on task T. That gives a great opportunity for applying the core decomposition in order to determine the best set A.

Algorithm 7. (k, η)-cores algorithm for core decomposition of probabilistic graphs.

Require: η
  for all $v \in V$ **do**
    compute the η-degree of v (Equation 20)
  end for
  $\mathbf{c} \leftarrow \emptyset, \mathbf{d} \leftarrow \emptyset, \mathbf{D} \leftarrow [\emptyset, \ldots, \emptyset]$
  for all $v \in V$ **do**
    $\mathbf{d}[v] \leftarrow \eta - deg(v)$
    $\mathbf{D}[\eta - deg(v)] \leftarrow \mathbf{D}[\eta - deg(v)] \cup \{v\}$
  end for
  for all $k = 0, 1, \ldots, n$ **do**
    while $\mathbf{D}[k] \neq \emptyset$ **do**
      $\mathbf{D}[k] \leftarrow \mathbf{D}[k] - \{v\}$, random $v \in \mathbf{D}[k]$
      $\mathbf{c}[v] \leftarrow k$
      for all $u : (u, v) \in E, \mathbf{d}[u] > k$ **do**
        recompute $\eta\text{-}deg(u)$
        $\mathbf{D}[\mathbf{d}[u]] \leftarrow \mathbf{D}[\mathbf{d}[u]] - \{u\}$
        $\mathbf{D}[\eta\text{-}deg(u)] \leftarrow \mathbf{D}[\eta\text{-}deg(u)] \cup \{u\}$
        $\mathbf{d}[u] \leftarrow \eta\text{-}deg(u)$
      end for
      $V \leftarrow V - \{v\}$
    end while
  end for
  return $\mathbf{c}$, n-dimensional vector containing the η-core number of each vertex in $\mathcal{G}$

Using the possible worlds model, the core decomposition concept can be applied to probabilistic graphs as well. Each vertex v has a probability of being a part of a k-core $\mathcal{H}$ which is defined as the probability that v has a degree greater or equal to k in $\mathcal{H}$. Then, a threshold η is applied to determine which vertices actually belong to the k-core based on this probability.

Therefore, given a probabilistic graph and a value for η, the (k, η)-core of $\mathcal{G}$ is defined as the maximal subgraph $\mathcal{H}$ such that the probability that each vertex belonging to it has a degree greater than or equal to k is greater than or equal to η. Thus, the (k, η)-core decomposition of a probabilistic graph is defined as the problem of finding the set of all (k, η)-cores of the graph. The η-degree of each vertex in $\mathcal{G}$ is defined as:

$$\eta - deg(v) = \max\{k \in [0, \ldots, d_v] | Pr[deg(v) \geq k] \geq \eta\} \qquad (20)$$

The outline of the (k, η)-core technique is outlined in Algorithm 7.

6 Distributed Mining of Uncertain Graphs

Having described some of the most important research contributions in uncertain graph mining, we center our focus on interesting issues related to uncertain graph mining in a distributed setting. Based on our previous discussion, it is evident that mining uncertain graphs is not trivial. In fact, it is considerably harder than mining conventional (i.e., certain) graphs. Therefore, efficient techniques are required to enable fast processing and provide meaningful results. A promising research direction with practical importance is the use of distributed engines that can utilize multiple resources (e.g., processors, memory, disks) aiming at mining massive datasets requiring significant space requirements. Nowadays, clusters running Spark or Hadoop are used consistently to provide the necessary functionality and performance.

Although many graph mining tasks have been parallelized for the case of conventional graphs, to the best of the authors' knowledge distributed algorithms for uncertain graphs have appeared only recently and for a limited set of problems. More specifically, in [9] the authors study reachability query processing in large uncertain graphs using distributed algorithms. However, there is a large set of mining tasks that can benefit from distributed computing. In the sequel we discuss briefly interesting problems related to uncertain graph mining in association with the corresponding difficulties that are raised due to the distributed setting.

Assume that the input data represents a single massive graph $G(V, E, p)$ with edge uncertainty. To facilitate distributed processing the first step is to split the input graph into several partitions. Partitions are distributed across cluster nodes to enable parallel execution of specific tasks. Graph partitioning [32] is an interesting problem on its own, and many algorithms have been proposed. For example, METIS [26] and FENNEL [55] are two very promising graph partitioning algorithms. Existing algorithms either work on unweighted or unweighted graphs. However, these techniques are not equipped with tools to handle uncertain graphs. Moreover, graph partitioning algorithms supported by distributed engines such as Spark, apply a hash-based approach where edges are distributed across machines in random order. However, many graph mining algorithms can benefit significantly by sophisticated partitioning algorithms. Therefore, there is a need for effective uncertain graph partitioning, taking into account the existential probabilities assigned to the edges of the graph. The way graph partitions are defined has a significant impact on mining tasks like node reachability and community detection.

Graph summarization [54] is a very interesting directions aiming at reducing the space requirements of massive graphs and at the same time keeping graph properties in order to provide answers to mining tasks using the summary. Graph summaries are either lossless (no information loss) or lossy (some information is lost having an impact on accuracy). To the best of the authors' knowledge the problem of uncertain graph summarization has not been addressed yet. Evidently, there is a need to compute graph summaries as fast as possible which means that distributed techniques may offer significant advantage over centralized approaches.

An important graph mining task is the discovery of the set of triangles. A triangle among nodes u, v and w is formed if all edges (u, v), (v, w) and (u, w) are present in the graph. It has been shown that triangles are important primitive structures that are essential in community formation [56] and many efficient algorithms have been proposed. To speed up processing, parallel algorithms for triangle discovery have been proposed [53]. However, in an uncertain graph it is natural to ask for the triangles with the maximum existential probability. In case of edge independence, the existential probability of a triangle is equal to the product of the probabilities of the corresponding edges. This corresponds to a top-k query, where k is the desired number of triangles. It is expected that distributed algorithms for top-k triangle discovery will offer significant performance improvement over centralized approaches.

Finally, a different set of problems may be defined in the case of a massive collection of small probabilistic graphs [40]. Instead of working with a single massive graph, a very large collection of small graphs is given as input. In such a case, there is a need to apply graph mining techniques across graphs. For example, clustering in this scenario involves the grouping of graphs in clusters, which means that meaningful similarity measures for graphs must be used. Moreover, the frequent pattern mining becomes an extremely difficult problem since not only we have to deal with *subgraph isomorphism*, but we are facing additional challenges due to the probabilistic nature of the graphs. It is expected that distributed techniques will have a significant impact on the performance of the algorithms that solve similar problems, since graphs may be distributed in several partitions and operations may be parallelized to enable more efficient execution of mining tasks.

7   Conclusions

In many real-life applications, graphs are annotated with existential probabilities on the edges, leading to the concept of uncertain or probabilistic graph. This means that each edge appears in the graph with a specific probability. Although this extension seems quite simple, it poses significant computational challenges in graph mining tasks that are easy to apply on conventional (deterministic) graphs.

The possible world semantics has proven to be a useful tool to mine probabilistic graphs, allowing efficient solutions to be developed in problems like graph clustering, clique discovery, k-nearest neighbor discovery and core decomposition. Although the possible world model implicates the existence and computation of $2^{|E|}$ deterministic graphs (in the worst case) for a single probabilistic graph $\mathcal{G} = (V, E, P)$, the algorithms proposed in the literature have been shown to produce effective results in their respective areas, while preserving the notion of uncertainty.

In this survey, we covered some of these methods focusing on the basic concepts and their associated algorithmic techniques. Taking into account that graph mining tasks applied to probabilistic graphs are characterized by increased complexity, it is natural to use distributed architectures towards more efficient processing. However, applying distributed processing to probabilistic graphs seems to be a challenging task by itself, and we argue that such techniques should be applied in order to facilitate faster execution times and to enable scalable data mining that could be applied to massive amounts of probabilistic graph data.

References

1. Aggarwal, C.C., Wang, H.: Managing and Mining Graph Data. Springer, Heidelberg (2010)
2. Ailon, N., Charikar, M., Newman, A.: Aggregating inconsistent information: ranking and clustering. J. ACM (JACM) **55**(5), 23 (2008)
3. Andersen, R., Chellapilla, K.: Finding dense subgraphs with size bounds. In: Avrachenkov, K., Donato, D., Litvak, N. (eds.) WAW 2009. LNCS, vol. 5427, pp. 25–37. Springer, Heidelberg (2009). doi:10.1007/978-3-540-95995-3_3
4. Bader, G.D., Hogue, C.W.: An automated method for finding molecular complexes in large protein interaction networks. BMC Bioinform. **4**(1), 2 (2003)
5. Bansal, N., Blum, A., Chawla, S.: Correlation clustering. Mach. Learn. **56**(1–3), 89–113 (2004)
6. Biswas, S., Morris, R.: Exor: opportunistic multi-hop routing for wireless networks. ACM SIGCOMM Comput. Commun. Rev. **35**(4), 133–144 (2005)
7. Bonchi, F., Gullo, F., Kaltenbrunner, A., Volkovich, Y.: Core decomposition of uncertain graphs. In: KDD, pp. 1316–1325 (2014)
8. Brandes, U., Gaertler, M., Wagner, D.: Engineering graph clustering: models and experimental evaluation. ACM J. Exp. Algorithmics **12**(1.1), 1–26 (2007)
9. Cheng, Y., Yuan, Y., Chen, L., Wang, G., Giraud-Carrier, C., Sun, Y.: Distr: a distributed method for the reachability query over large uncertain graphs. IEEE Trans. Parallel Distrib. Syst. **27**(11), 3172–3185 (2016)
10. Colbourn, C.J., Colbourn, C.: The Combinatorics of Network Reliability, vol. 200. Oxford University Press, New York (1987)
11. Cook, D.J., Holder, L.B.: Mining Graph Data. Wiley, Hoboken (2006)
12. Eppstein, D., Löffler, M., Strash, D.: Listing all maximal cliques in sparse graphs in near-optimal time. In: Cheong, O., Chwa, K.-Y., Park, K. (eds.) ISAAC 2010. LNCS, vol. 6506, pp. 403–414. Springer, Heidelberg (2010). doi:10.1007/978-3-642-17517-6_36
13. Feo, T.A., Resende, M.G.: A probabilistic heuristic for a computationally difficult set covering problem. Oper. Res. Lett. **8**(2), 67–71 (1989)

14. Fogaras, D., Rácz, B.: Towards scaling fully personalized pagerank. In: Leonardi, S. (ed.) WAW 2004. LNCS, vol. 3243, pp. 105–117. Springer, Heidelberg (2004). doi:10.1007/978-3-540-30216-2_9

15. Fortunato, S.: Community detection in graphs. Phys. Rep. **483**(3), 75–174 (2010)

16. Friden, C., Hertz, A., de Werra, D.: Stabulus: a technique for finding stable sets in large graphs with tabu search. Computing **42**(1), 35–44 (1989)

17. Gavin, A.-C., Bösche, M., Krause, R., Grandi, P., Marzioch, M., Bauer, A., Schultz, J., Rick, J.M., Michon, A.-M., Cruciat, C.-M., et al.: Functional organization of the yeast proteome by systematic analysis of protein complexes. Nature **415**(6868), 141–147 (2002)

18. Ghosh, J., Ngo, H.Q., Yoon, S., Qiao, C.: On a routing problem within probabilistic graphs and its application to intermittently connected networks. In: 26th IEEE International Conference on Computer Communications, INFOCOM 2007, pp. 1721–1729. IEEE (2007)

19. Gionis, A., Mannila, H., Tsaparas, P.: Clustering aggregation. ACM Trans. Knowl. Discov. Data (TKDD) **1**(1), 4 (2007)

20. Glover, F.: Tabu search–part II. ORSA J. Comput. **2**(1), 4–32 (1990)

21. Goyal, A., Lu, W., Lakshmanan, L.V.: CELF++: optimizing the greedy algorithm for influence maximization in social networks. In: Proceedings of the 20th International Conference Companion on World Wide Web, pp. 47–48. ACM (2011)

22. Harley, E., Bonner, A., Goodman, N.: Uniform integration of genome mapping data using intersection graphs. Bioinformatics **17**(6), 487–494 (2001)

23. Huang, X., Cheng, H., Yu, J.X.: Attributed community analysis: global and ego-centric views. Data Eng. **14**, 29 (2016)

24. Huang, X., Lu, W., Lakshmanan, L.V.: Truss decomposition of probabilistic graphs: semantics and algorithms. In: SIGMOD, pp. 77–90 (2016)

25. Jin, R., Liu, L., Aggarwal, C., Shen, Y.: Reliable clustering on uncertain graphs. In: ICDM, pp. 459–468 (2012)

26. Karypis, G., Kumar, V.: Parallel multilevel k-way partitioning scheme for irregular graphs. In: Proceedings of the 1996 ACM/IEEE Conference on Supercomputing, Supercomputing 1996, Washington, DC, USA. IEEE Computer Society (1996)

27. Khan, A., Bonchi, F., Gionis, A., Gullo, F.: Fast reliability search in uncertain graphs. In: EDBT, pp. 535–546 (2014)

28. Kollios, G., Potamias, M., Terzi, E.: Clustering large probabilistic graphs. IEEE Trans. Knowl. Data Eng. **25**(2), 325–336 (2013)

29. Kortsarz, G., Peleg, D.: Generating sparse 2-spanners. J. Algorithms **17**(2), 222–236 (1994)

30. Kovács, F., Legány, C., Babos, A.: Cluster validity measurement techniques. In: 6th International Symposium of Hungarian Researchers on Computational Intelligence. Citeseer (2005)

31. Krogan, N.J., Cagney, G., Yu, H., Zhong, G., Guo, X., Ignatchenko, A., Li, J., Pu, S., Datta, N., Tikuisis, A.P., et al.: Global landscape of protein complexes in the yeast Saccharomyces cerevisiae. Nature **440**(7084), 637–643 (2006)

32. LaSalle, D., Patwary, M.M.A., Satish, N., Sundaram, N., Dubey, P., Karypis, G.: Improving graph partitioning for modern graphs and architectures. In: Proceedings of the 5th Workshop on Irregular Applications: Architectures and Algorithms, IA3 2015, pp. 14:1–14:4. ACM, New York (2015)

33. Liu, L., Jin, R., Aggarwal, C., Shen, Y.: Reliable clustering on uncertain graphs. In: 2012 IEEE 12th International Conference on Data Mining (ICDM), pp. 459–468. IEEE (2012)

34. Mcauley, J., Leskovec, J.: Discovering social circles in ego networks. ACM Trans. Knowl. Discov. Data (TKDD) **8**(1), 4 (2014)
35. Mewes, H.-W., Amid, C., Arnold, R., Frishman, D., Güldener, U., Mannhaupt, G., Münsterkötter, M., Pagel, P., Strack, N., Stümpflen, V., et al.: MIPS: analysis and annotation of proteins from whole genomes. Nucleic Acids Res. **32**(suppl 1), D41–D44 (2004)
36. Mukherjee, A., Xu, P., Tirthapura, S.: Enumeration of maximal cliques from an uncertain graph. IEEE Trans. Knowl. Data Eng. **29**, 543–555 (2016)
37. Mukherjee, A.P., Xu, P., Tirthapura, S.: Mining maximal cliques from an uncertain graph. In: 2015 IEEE 31st International Conference on Data Engineering (ICDE), pp. 243–254. IEEE (2015)
38. Newman, M.E.: Modularity and community structure in networks. Proc. Nat. Acad. Sci. **103**(23), 8577–8582 (2006)
39. Palla, G., Derényi, I., Farkas, I., Vicsek, T.: Uncovering the overlapping community structure of complex networks in nature and society. Nature **435**(7043), 814–818 (2005)
40. Papapetrou, O., Ioannou, E., Skoutas, D.: Efficient discovery of frequent subgraph patterns in uncertain graph databases. In: Proceedings of EDBT, pp. 355–366 (2011)
41. Parchas, P., Gullo, F., Papadias, D., Bonchi, F.: The pursuit of a good possible world: extracting representative instances of uncertain graphs. In: SIGMOD, pp. 967–978 (2014)
42. Parchas, P., Gullo, F., Papadias, D., Bonchi, F.: Uncertain graph processing through representative instances. ACM Trans. Database Syst. **40**(3), 20:1–20:39 (2015)
43. Pathak, N., Mane, S., Srivastava, J.: Who thinks who knows who? Socio-cognitive analysis of email networks. In: Sixth International Conference on Data Mining, ICDM 2006, pp. 466–477. IEEE (2006)
44. Pattillo, J., Youssef, N., Butenko, S.: Clique relaxation models in social network analysis. In: Thai, M.T., Pardalos, P.M. (eds.) Handbook of Optimization in Complex Networks. Springer Optimization and Its Applications, vol. 58, pp. 143–162. Springer, New York (2012)
45. Pfeiffer, J., Neville, J.: Methods to determine node centrality and clustering in graphs with uncertain structure. In: ICWSM (2011)
46. Potamias, M., Bonchi, F., Gionis, A., Kollios, G.: K-nearest neighbors in uncertain graphs. Proc. VLDB Endow. **3**, 997–1008 (2010)
47. Rokhlenko, O., Wexler, Y., Yakhini, Z.: Similarities and differences of gene expression in yeast stress conditions. Bioinformatics **23**(2), e184–e190 (2007)
48. Rysz, M., Mirghorbani, M., Krokhmal, P., Pasiliao, E.L.: On risk-averse maximum weighted subgraph problems. J. Comb. Optim. **28**(1), 167–185 (2014)
49. Sarkar, P., Moore, A.W., Prakash, A.: Fast incremental proximity search in large graphs. In: Proceedings of the 25th International Conference on Machine Learning, pp. 896–903. ACM (2008)
50. Seidman, S.B.: Network structure and minimum degree. Soci. Netw. **5**(3), 269–287 (1983)
51. Sevon, P., Eronen, L., Hintsanen, P., Kulovesi, K., Toivonen, H.: Link discovery in graphs derived from biological databases. In: Leser, U., Naumann, F., Eckman, B. (eds.) DILS 2006. LNCS, vol. 4075, pp. 35–49. Springer, Heidelberg (2006). doi:10.1007/11799511_5
52. Shamir, R., Sharan, R., Tsur, D.: Cluster graph modification problems. Discrete Appl. Math. **144**(1), 173–182 (2004)

53. Tangwongsan, K., Pavan, A., Tirthapura, S.: Parallel triangle counting in massive streaming graphs. In: Proceedings of the 22nd ACM International Conference on Information & Knowledge Management, CIKM 2013, New York, NY, USA, pp. 781–786. ACM (2013)

54. Tian, Y., Hankins, R.A., Patel, J.M.: Efficient aggregation for graph summarization. In: Proceedings of the 2008 ACM SIGMOD International Conference on Management of Data, SIGMOD 2008, New York, NY, USA, pp. 567–580. ACM (2008)

55. Tsourakakis, C., Gkantsidis, C., Radunovic, B., Vojnovic, M.: Fennel: streaming graph partitioning for massive scale graphs. In: Proceedings of the 7th ACM International Conference on Web Search and Data Mining, WSDM 2014, New York, NY, USA, pp. 333–342. ACM (2014)

56. Tsourakakis, C.E.: A novel approach to finding near-cliques: the triangle-densest subgraph problem. CoRR abs/1405.1477 (2014)

57. Valiant, L.G.: The complexity of enumeration and reliability problems. SIAM J. Comput. **8**(3), 410–421 (1979)

58. Wu, Y., Yang, Y., Jiang, F., Jin, S., Xu, J.: Coritivity-based influence maximization in social networks. Phys. A Stat. Mech. Appl. **416**, 467–480 (2014)

59. Yezerska, O., Butenko, S., Boginski, V.L.: Detecting robust cliques in graphs subject to uncertain edge failures. Ann. Oper. Res. **238**, 1–24 (2016)

60. Yuan, Y., Wang, G., Chen, L., Wang, H.: Efficient subgraph similarity search on large probabilistic graph databases. Proc. VLDB Endow. **5**, 800–811 (2012)

61. Zhang, B., Park, B.-H., Karpinets, T., Samatova, N.F.: From pull-down data to protein interaction networks and complexes with biological relevance. Bioinformatics **24**(7), 979–986 (2008)

62. Zou, Z.: Polynomial-time algorithm for finding densest subgraphs in uncertain graphs. In: Proceedings of MLG Workshop (2013)

63. Zou, Z., Li, J., Gao, H., Zhang, S.: Finding top-k maximal cliques in an uncertain graph. In: 2010 IEEE 26th International Conference on Data Engineering (ICDE), pp. 649–652. IEEE (2010)

64. Zou, Z., Li, J., Gao, H., Zhang, S.: Mining frequent subgraph patterns from uncertain graph data. IEEE Trans. Knowl. Data Eng. **22**(9), 1203–1218 (2010)

65. Zou, Z., Zhu, R.: Truss decomposition of uncertain graphs. Knowl. Inf. Syst. **50**, 197–230 (2016)

Recovering from Cloud Application Deployment Failures Through Re-execution

Ioannis Giannakopoulos[1]([⊠]), Ioannis Konstantinou[1], Dimitrios Tsoumakos[2], and Nectarios Koziris[1]

[1] Computing Systems Laboratory, National Technical University of Athens, Athens, Greece
{ggian,ikons,nkoziris}@cslab.ece.ntua.gr
[2] Department of Informatics, Ionian University, Corfu, Greece
dtsouma@ionio.gr

Abstract. In this paper we study the problem of automated cloud application deployment and configuration. Transient failures are commonly found in current cloud infrastructures, attributed to the complexity of the software and hardware stacks utilized. These errors affect cloud application deployment, forcing the users to manually check and intervene in the deployment process. To address this challenge, we propose a simple yet powerful deployment methodology with error recovery features that bases its functionality on identifying the script dependencies and re-executing the appropriate configuration scripts. To guarantee the idempotent script execution, we adopt a filesystem snapshot mechanism that enables our approach to revert to a healthy filesystem state in case of failed script executions. Our experimental analysis indicates that our approach can resolve any transient deployment failure appearing during the deployment phase, even in highly unpredictable cloud environments.

1 Introduction

The evolution of cloud computing has aroused new needs regarding task automation, i.e., the automatic execution of complex tasks without needing human intervention, especially in the field of application deployment. The users need to be able to describe their applications, deploy them over the cloud infrastructure of their choice without needing to execute complex tasks such as resource allocation, software configuration, etc., manually. Through automation, they can instantly build complex environments fully or semi automatically [18], enhance portability through which they can easily migrate their applications into different cloud providers, facilitate the deployment of complex applications, by decomposing the complex description into simpler and easier to debug components, etc. Nevertheless, automated resource orchestration and software configuration practices act as a prerequisite towards cloud elasticity [25], which dictates that resources and software modules must be configured automatically as in [28].

To automate the application deployment, a number of systems and tools has been proposed such as Heat [13], Sahara [14], Juju [12], CloudFormation [3],

© Springer International Publishing AG 2017
T. Sellis and K. Oikonomou (Eds.): ALGOCLOUD 2016, LNCS 10230, pp. 117–130, 2017.
DOI: 10.1007/978-3-319-57045-7_7

BeanStalk [4], Wrangler [21], etc. The behavior of these systems is similar: They first communicate with the cloud provider so as to allocate the necessary resources and then execute configuration scripts in order to deploy the software components into the newly allocated resources. The configuration process can be facilitated through the utilization of popular configuration tools such as Chef [8], Puppet [16] and Ansible [1]. Since each script may depend on multiple different components (resources and software modules), each of the aforementioned systems creates a dependency graph in order to execute the configuration actions in the correct order and synchronize the concurrently executed configuration tasks.

A shortcoming of the aforementioned approaches, though, is that they do not take into consideration the dynamic and, sometimes, unstable nature of the cloud. Services downtime due to power outages, hardware failures, etc. [5], unpredictable boot times [6] closely related to the provider's load are some of the factors that contribute to the instability of the cloud. Furthermore, unpredictability is exaggerated by the fact that the current cloud infrastructures have increased the complexity of the utilized software and hardware stacks [20], something that contributes to the appearance of *transient* failures, such as network glitches [26], host reboots, etc. These failures are short but can produce severe service failures [11]. Furthermore, since the application deployment demands cooperation and synchronization between multiple parties including the cloud provider, the VM's services and other external parties, it becomes apparent that these transient errors are commonly found and can lead application deployment to failure, leaving, in the worst case, stale resources that need manual handling, e.g., a partially deployed application may have successfully created some VMs or volumes that need to be deleted before triggering a new deployment.

To tackle the aforementioned challenges, we propose an application deployment methodology which is able to recover from transient infrastructure errors through re-execution of the necessary scripts. Specifically, the deployment is modeled as a set of scripts concurrently executed for different software modules and exchanging messages, forming a directed acyclic graph. If a script execution fails, our approach traverses the graph, identifies the scripts that need to be re-executed and executes them. Since a script execution may be accompanied with implicit side effects, we adopt a filesystem snapshot mechanism similar to the mechanism used by Docker [9] which guarantees that a script execution remains idempotent for all the filesystem-related resources, enabling our approach to re-execute a script as many times as needed for the deployment to complete. Through an extensive evaluation for complex deployment graphs, we showcase that our approach remains effective even in unpredictable cloud environments. Our contributions can be, thus, summarized as follows:

- we present an error recovery methodology that can efficiently restart partially completed application deployments,
- we provide a powerful methodology, that allows to snapshot the filesystem and easily revert into a previous state, in case of failure,

– we demonstrate that our approach achieves the deployment of complex applications in times comparable to the failure-free cases, even when the error rate
is extremely high.

2   Deployment and Configuration Model

Assume a typical three-tier application consisting of the following modules: a
Web Server (rendering and serving Web Pages), an Application Server (implementing the business logic) and a Database Server (for the application data). For
simplicity's sake, we assume that each module runs in a dedicated server and the
application will be deployed in a cloud provider. If the application deployment
occurred manually, one should create three VMs and connect (e.g., via SSH)
to them in order to execute scripts that take care of the configuration of the
software modules. In many cases, the scripts need input that is not available
prior to the resource allocation. For example, the Application Server needs to
know the IP address and the credentials of the Database Server in order to be
able to establish a connection to it and function properly. In such cases, the
administrator should manually provide this *dynamic* information.

In order to automate the deployment and configuration, we need to create a communication channel between different application modules in order to
exchange messages containing such dynamic information. Such a channel can be
implemented via a simple queueing mechanism. Each module publishes information needed by the rest of the modules and subscribes to queues, consuming
messages produced by other modules. The deployment scripts are executed up to
a point where they expect input from another module. At these points, they
block until the expected input is received, i.e., a message is sent from another
module and it is successfully transmitted to the blocked module. The message
transmission is equivalent to posting a new message into the queue (from the
sender's perspective) and consuming this message from the queue (from the
receiver's perspective). In cases that no input is received (after the expiration of
time threshold), the error recovery mechanism is enabled. To provide a better
illustration of the deployment process, consider Fig. 1. In this Figure, a deployment process between two software modules named (1) and (2) is depicted. The
vertical direction represents the elapsed time and the horizontal arrows represent message exchange[1]. At first, both (1) and (2) begin the deployment process
until points A and A' are reached respectively. When (1) reaches A, it sends a
message to (2) and proceeds. On the other side, (2) blocks at point A' until the
message sent from (1) arrives and, upon arrival, consumes it and continues with
the rest of the script. If the message does not arrive, then the recovery procedure
is triggered, as described in Sect. 3.

From the above description it is obvious that when a module needs to send
a message, there is no need to wait until this message is successfully received by

---

[1] Note that message transmission might not be instant (as implied by the Figure)
since consumption of a specific message might occur much later than the message
post, but the arrows are depicted perpendicular to the time axis for simplicity.

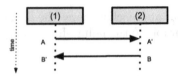

Fig. 1. Message exchange between modules

the other module. However, in some cases, this blocking send may be desired. For example, take the case where two modules negotiate about a value, e.g., a randomly generated password assigned to the root account of the Database Server. Assume that module (1) (the Application Server) decides on the value and sends it to module (2) (the Database Server). Module (1) must ensure that the password is set, prior to trying to connect to the database. To this end, (2) can send an acknowledgment as depicted between points B and B'. In this context, it becomes apparent that the message exchange protocol can also function as a synchronization mechanism. This scheme represents a dependency graph between the application's modules, since each incoming horizontal edge (e.g., the one entering at point A') declares that the execution of a configuration script depends on the correct execution of another. Schemes like the one presented at Fig. 1 present a circular dependency, since both modules (1) and (2) depend on each other, but on different time steps. Various state-of-the-art deployment mechanisms do not handle circular dependencies (e.g., Openstack Heat [13]) since they do not support such a message exchange mechanism during the application configuration. Furthermore, such a circularity could easily lead to deadlocks, i.e., a case in which all the modules are blocked because the wait input from another (also blocked) module. We assume that the user is responsible for creating deadlock-free deployment descriptions.

This message exchange mechanism can easily be generalized so as to be applied when an elasticity action is applied, i.e., a dynamic resource allocation during the application's lifetime. A simple *elastic* action can be decomposed into three steps: (a) the preparation step, where scripts are executed in order to prepare the application for a resizing action, (b) the IaaS communication step, so as to allocate/deallocate resources and (c) the post action step, where the application is configured in order to function properly after the IaaS action. Steps (a) and (c) can be expressed with a similar messaging/synchronization mechanism described before; Step (b), on the other hand, entails the communication with the cloud provider to orchestrate the application in the infrastructure level. Through the composition of such resizing actions, it is possible to express a composite resizing action affecting many application modules in parallel. Finally, the same mechanism is applied in cases where a module consists of multiple module instances (e.g., different nodes of a NoSQL cluster). In these cases, each module instance is addressed as a new entity and interacts with the rest modules separately.

3   Error Detection and Recovery

We now describe the mechanism through which the deployment errors are identified. During the deployment process, a module instance may send a message to another module. This message may contain information needed by the later module in order to proceed with the its deployment, or it could just be a synchronization point. In any case, the second module blocks the deployment execution until the message arrives, whereas the first module sends its message and proceeds with the execution of its deployment script. In the case of an error, a module will not be able to send a message and the receiver will remain blocked. To this end, we set a timeout period that, if exceeded, the waiting module broadcasts messages to the other modules informing them that an error has occurred. From that point on, the error recovery mechanism takes control, evaluates the error, as we will shortly describe, and performs the necessary actions in order to restore the deployment procedure. Many modules may only consume messages and not create any. In such cases, a possible failure is identified with the same mechanism in the final part of the deployment. When a module finalizes its deployment, it broadcasts a message informing the rest modules that the process is finished. When all the modules receive these messages from the rest modules, the process is considered finished. Eventually, even when a module does not need to send a message, a possible failure of its execution will be identified at the final state of its execution, since it will not send a termination acknowledgment.

Error recovery is based on the repetition of the execution of scripts/actions which led the system to a failure. Given the previously described error identification mechanism, at some random point a module stalls waiting for a message from another module. It then broadcasts a special message informing the other modules that the deployment may have failed. A *master* node is then elected (among the existing modules) and this node undertakes the responsibility to execute the health check algorithm. When it identifies that a script execution has failed, it *backtracks* and informs the responsible module that it should repeat the execution of the scripts before the failure. For example, assume the more complex example provided in Fig. 2.

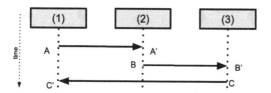

Fig. 2. More complex deployment example

Assume that an error is identified in B', meaning that (2) did not send a message to (3). Since (2) had received a message from (1) in A', this means that the problematic script is the one executed between A' and B. Let's repeat the same example now, but assume that the problem is now found in C'. After point A, (1) will remain blocked until (3) sends a message when point C is reached. If the timeout is exceeded, then (1) triggers the Health Check mechanism. In that case the master (any of (1), (2) or (3)) module will backtrack until all the dependencies for each state are resolved, and finally identify whether a failure occurred or the scripts needed more time to finish. If a problem did occur, the master node informs the responsible module and this module, accordingly, re-executes the necessary scripts. From the above analysis, it becomes apparent that the absence of a message is indicative of a possible failure. The reason behind this absence, though, remains obscure: A script might have finished its execution and attempted to send a message but the message never reached the communication channel due to unreliable network. Our approach views all the possible alternatives in a unified manner since we make the assumption that the communication channel is also unreliable and address the channel's errors as deployment errors.

In order to identify the dependencies between the different software modules, we address the problem of error recovery as a graph traversal problem. The idea is that the messaging/synchronization scheme is translated in a graph representing the dependencies between different modules. For example, the deployment represented in Fig. 2 can be translated as shown in Fig. 3(a). The arrows in this Figure represent either a script execution (vertical edges) or a message exchange (horizontal edges). The top nodes represent the state of each VM exactly after the VM bootstrapping has finished. The bottom graph nodes represent the final state of the deployment for each module. The intermediate nodes correspond to intermediate states of the modules, where messages are sent and consumed. At the end of the process, all the modules exchange messages once more, to verify that the process is successfully finished. Since receiving and sending messages in the same module may lead to a deadlock, we serialize the modules' actions and enforce them to first send any messages and then block to receive them.

Assume now that, at some point, an error occurs. The master node first parses the deployment graph as presented in Fig. 3(a) and creates the dependency graph, as depicted in Fig. 3(b), which is essentially the deployment graph

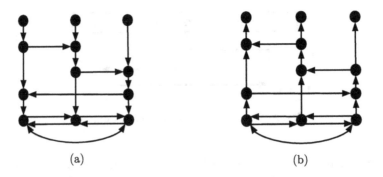

(a)   (b)

Fig. 3. Deployment and dependency graph of Fig. 2

with inverted edges. Then, it executes Algorithm 1 where the dependency graph is traversed in a Breadth First order, starting from the node where the failure was detected and the healthy states are, then, identified. Specifically, if a visited node is not healthy, then the graph traversal continues to its children, else the traversal stops. In this context, *healthy* states are considered to be the intermediate configuration states (nodes on the graph) that have been reached by the deployment execution without an error. This way, the algorithm manages to identify the most recent healthy intermediate states reached by each module. With this information, one can easily identify which steps should be repeated.

Algorithm 1. Health Check Algorithm

Require: transpose deployment graph T, failed node n
Ensure: list of *healthy* nodes $frontier$
1: $failed \leftarrow \{n\}$
2: $frontier \leftarrow \emptyset$
3: **while** $failed \neq \emptyset$ **do**
4:     $v = \text{pop}(failed)$
5:     **for** $t \in \text{neighbors}(T, v)$ **do**
6:         **if** failed(t) **then**
7:             $failed \leftarrow failed \cup \{t\}$
8:         **else**
9:             $frontier \leftarrow frontier \cup \{t\}$
10:         **end if**
11:     **end for**
12: **end while**
13: **return** $frontier$

In various cases, script re-execution means that a module must publish a new message and its content may be different for each execution. For example, if a module creates a random password and broadcast it to the rest modules, each script re-execution generates a new password. In other cases though, the content of the message might be the same: e.g., when a module sends the IP address of the VM that is hosted into, the content of the message remains the same since it is independent from the script execution. In the former case, the modules that depend on the script-dependent messages should be re-executed as well. For example, if the Application Server sends a new password to the Database Server, this means that the Database Server should be reconfigured so as to reflect this change (even if the Database Server presented no errors), otherwise the Application Server will not be able to establish a connection with the later. Our approach views all the messages that are exchanged between different modules as script-dependent and forces the re-execution for each module that depends on a specified message. The script-independent messages should be transmitted at the first steps of the deployment so as to avoid pointless script re-execution. Such a practice is also followed by other popular deployment tools such as Openstack

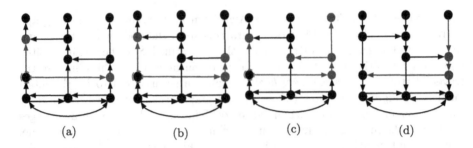

Fig. 4. Example of Health Check algorithm (Color figure online)

Heat, since all the script-independent information is considered known to every software module, prior to the execution of the configuration scripts.

To further clarify the algorithm's execution, we provide an example in which we demonstrate the way that the Health Check algorithm traverses the dependency graph when an error occurs in Fig. 4. Figure 4(a), (b) and (c) demonstrate the iterations of the algorithm, until the healthy nodes (green nodes) are identified. Figure 4(d) demonstrates the scripts that are re-executed so as to continue with the deployment (identified as blue edges). Two notes must be made at this point. First, the incoming vertical edge to the node in which the error was identified need not be re-executed. This is attributed to the fact that the script execution must have been successful, else the module would never have reached to the point where it expects a message, thus the Health Check mechanism would not have been triggered. Second, it is obvious that script-dependent messages need to be resent and demand script re-execution only when they are sent and not when they are consumed by a script. When a module need to re-consume an message generated by another module, we assume that the message is already available through the communication channel and the sender does not need to resend any information.

4   Idempotent Script Execution

The idea of script re-execution when a deployment failure occurs, is only effective when two important preconditions are met: (a) the failures are transient and (b) if a script is executed multiple times it always leads the deployment into the same state, or, simpler, it always has the same side effects, i.e., it is *idempotent*. In this, section we discuss these preconditions and describe how are these enforced through our approach.

First, a failure is called "transient" when it is caused by a cloud service and it was apparent for a short period of time. Network glitches, routers unavailability due to a host reboot and network packet loss are typical examples of such failures caused by the cloud platform but, in most cases, they are only apparent for a short time (seconds or a few minutes) and when disappeared the infrastructure is completely functional. Various works study those types of failures [26] and attribute them to the complexity of the cloud software and hardware stack. Although most cloud providers protect their customers from such failures through their SLAs [17], they openly discuss them and provide instruction for failures caused by sporadic host maintenance tasks [7] and various other reasons [19]. Since the cloud environment is so dynamic and the automation demanded by the cloud orchestration tools requires the coordination of different parties, script re-execution is suggested on the basis that such transient failures will eventually disappear and the script will become effective.

However, in the general case, if a script is executed multiple times it will not always have the same effect. For example, take the case of a simple script that reads a specific file from the filesystem and deletes it. The script can only be executed exactly once; The second time it will be executed it will fail since the file is no longer available. This failure is caused by the side effects (the file deletion) caused by the script execution, which lead the deployment to a state in which the same execution cannot be repeated. To overcome this limitation, we adopt a *snapshot* mechanism to capture the VM's filesystem prior to script execution, allow the script to be executed and in case of failure revert it to the previous healthy state. This mechanism is based on building the VM's filesystem through the composition of different layers. The base layer consist of the VM's filesystem. Any layer on top of this contains only the updated files along with their new content. The *revert* action is equivalent to removing the top layer of the filesystem and the script's side effects are vanished. To ensure the idempotent property, we generate a new layer prior to each script execution. This way, the side effects of each script are "undoable" and, in case of failure, we guarantee that the VM's filesystem will be identical to the one before the failed script execution.

From a technical perspective, there exists various solutions which can be utilized to achieve this snapshot mechanism. In our approach, we utilize *AUFS* (Another Union FileSystem) [2], which is a special case of a Union Filesystem. AUFS allows different directories to be mounted on the same mountpoint where each directory represents a layer. The ordering of the directories determines the order of the layers into the filesystem. The addition of a new layer or the removal of an unwanted layer is equivalent to creating and removing a directory respectively. Different solutions such as overlayfs [15] and btrfs [27] could also be utilized, but all these solutions are equivalent in terms of practicality and only differ in terms of available features and performance. Since this snapshot mechanism is adopted by Docker [9], a popular Linux Container implementation, we utilized AUFS since it is considered to be the most stable and achieves the highest performance among the proposed solutions [10]. Finally, we must note that through the previously described mechanism we guarantee that only filesystem-related actions remain idempotent. Any actions that have side effects towards non disk based or external resources, e.g., external API calls, memory state updates, etc., cannot be addressed by our approach and, thus, cannot be reverted.

5   Experimental Evaluation

A deployment can be expressed in multiple different ways for one application and the execution time can vary greatly according to the used scripts. We test our methodology for various deployment graphs, as the one depicted in Fig. 3, for different sizes (number of nodes and edges) and different failure probabilities. The number of vertical paths represents the number of modules (or module instances). The number of states inside a vertical path is expressed as the number of *levels* of synchronization. For each graph, we create a random number of messages, i.e., two nodes belonging to the same level of different module are connected with an edge with a specified probability. This measure expresses the cohesion of the graph that is defined as the degree of dependency between different modules. Finally, during the deployment each vertical edge (script execution) may fail with a specific probability. We will study how these parameters affect the duration of the deployment process measured not in absolute time, since time is dependent to the duration of each resizing script and it greatly varies between different applications, but in number of edges.

Each module begins from an initial node of the deployment graph (for each module) and terminates in a terminal node (for the respective module). The longest path traversed, is equal to the duration of the deployments. Each module stalls when an error occurs, as described in the previous sections. We divide this path with the number of steps needed for the deployment in the optimal case, that is if no failures occurred. The results are presented in Fig. 5. The left most Figure describes the relationship between the Error Rate of the scripts and the Relative Execution time for increasing number of deployment levels, whereas the right Figure depicts the same relationship for an increasing number of application modules. The cohesion factor defined as the possibility for one module to have an outgoing edge in a specific node is set to 0.2. For each graph size we created 100 random graphs and for each graph we executed the deployment algorithm 5 times. The graphs depict the averages of those runs.

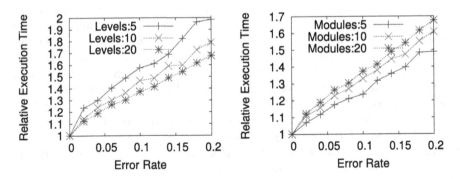

Fig. 5. Relative Execution Time vs Error Rate

In both graphs it is obvious that an increase in Error Rate leads to increased Relative Execution Time. The left graph shows that deployments described with less steps per module are higher affected by an increase in the Error Rate than the ones with a large number of levels per modules. More levels lead to execution of more scripts, thus more scripts will probably fail. However, each failure costs less, since the repetition of one script will traverse a smaller portion of the graph and eventually, as seen from the comparison with the optimal case, the overall overhead inserted by the re execution declines with the number of levels. The deployments depicted in this figure consist of 20 modules. On the contrary, applications that consist of many modules are affected more by an increase in the Error Rate (on the rightmost Figure). The five modules case is deployed 1.5 slower than the optimal case (when the error rate is 0.2), while simultaneously, the 20 modules case is deployed 1.7 times slower. This means that when the number of modules quadruples, the increase in Relative Execution Time is less than 15%, making our methodology suitable for complex applications, consisting of multiple modules and many synchronization levels per module. Comparing with the traditional deployment tools, where no error recovery mechanisms are employed, it is obvious that the termination of the deployment occurs orders of magnitude faster, since the probability for a successful deployment is $(1 - ER)^{(M \cdot N)}$, where ER is the Error Rate, M equals the number of application modules and N is the number of deployment levels for each module. This quantity grows exponentially with the application's complexity making these tools unsuitable for complex applications deployed in unpredictable environments.

6   Related Work

Automated cloud application deployment and configuration is a widespread problem, evolving with the expansion of cloud computing. Various systems, both in industry and academia, have been proposed in order to handle application deployment and scaling. In the Openstack ecosystem, Heat [13] and Sahara [14] have been proposed. Heat is a generic deployment tool used to define and configure deployment *stacks* that involve different resources (e.g., VMs, volumes, networks, IP addresses, etc.) and target to automate their management and orchestration. Heat retains a simple deployment model where it assumes that any script-independent information is available to each VM prior to the script execution and it does not support cyclical dependencies between different application modules. It also retains autoscaling capabilities, by monitoring the allocated resources and performing simple rule based actions in order to scale the running applications. Sahara (the ex Savannah project) is targeted to the deployment and configuration of Data Processing systems, such as Hadoop and Spark. Both tools integrate solely with Openstack and do not provide error recovery features.

The AWS counterparts of these tools are [3,4]. CloudFormation is a generic tool, relevant to Openstack Heat and it retains a similar deployment model where the user defines the different resources that should be allocated along with configuration scripts that are responsible for the software configuration. Elastic BeanStalk is primarily used for deploying specific applications (e.g., Web Servers, Load Balancers, etc.) into the Amazon cloud and also provide elastic capabilities. Finally, Canonical Juju [12] is another system used for deployment and scaling. Different modules are organized as *charms*. A charm is, essentially, a set of configuration files that determine the way through which a software module is configured and deployed and how it interacts with other modules. The user can choose from a set of preconfigured charms and compose complex applications structures. All the aforementioned systems target to provide a level of automation through which a user can easily deploy and configure their application. However, the dynamic cloud nature is not taken into consideration since in case of failure the deployment is aborted and the user must manually resume it or cancel it. On the contrary, our approach allows the re-execution of failed scripts so as to automatically resume the deployment.

Similar to the previous systems, there exist standalone tools specializing in creating an identical application environment, mainly for development reasons. The most representative is [18]. It cooperates with popular configuration management tools, such as [1,8,16] and creates an identical virtualized application environment, supporting multiple hypervisors and cloud providers. Runtime orchestration and scaling are not considered, though. Furthermore, in [21] *Wrangler* is proposed, a system that bases its functionality in script executions (called plugins). The philosophy of this system is similar to our approach; However error recovery is not considered. In [23], a data-centric approach is presented formulating resources as transactions, exploiting their ACID properties for error recovery. However, each action should have an "undo" action in order to participate in a transaction, which is a stronger hypothesis than the idempotent property enforced by the layered filesystems proposed in this work. In [22] a synchronization framework for Chef [8] is introduced: The user can introduced "breakpoints" into the execution of Chef recipes so as to synchronize the application configuration. This work identifies the need for synchronization when a cloud application is deployed; However, it does not handle transient deployment failures. Finally, in [24] an approach through which the cloud APIs are overridden in order to obtain knowledge about the status of their commands. This work is not focused on the application configuration as the current work is; However, it identifies the error-prone nature of the synchronous cloud providers and suggests a solution for transforming the unreliable cloud APIs into reliable calls with predictable outputs.

7   Conclusions

In this paper, we proposed a cloud deployment and configuration methodology, capable to overcome transient cloud failures through re-executing the failed deployment scripts. Our methodology resolves the dependencies among different software modules and identifies the scripts that should be re-executed so as to resume the application deployment. To guarantee that each script will have the same effect, we utilize a layered filesystem architecture through which we snapshot the filesystem before the script execution and, in case of failure, revert it to the previous state and retry. Our evaluation indicates that our methodology is effective even for the most unstable cloud environments and it is particularly effective for application that consist of many different modules exchanging multiple messages throughout the process.

References

1. Ansible. http://www.ansible.com/home
2. AUFS. http://aufs.sourceforge.net/
3. AWS CloudFormation. http://aws.amazon.com/cloudformation/
4. AWS Elastic BeanStalk. http://aws.amazon.com/elasticbeanstalk/
5. AWS Incident. https://goo.gl/f959fl
6. AWS Instances Boot Times. http://goo.gl/NQ1qNw
7. AWS Maintenance. https://aws.amazon.com/maintenance-help/
8. Chef. https://www.chef.io/chef/
9. Docker Container. https://www.docker.com/
10. Docker: Select a storage driver. https://goo.gl/o383To
11. Google App Engine Incident. https://goo.gl/ICI0Mo
12. Juju. https://juju.ubuntu.com/
13. Openstack Heat. https://wiki.openstack.org/wiki/Heat
14. Openstack Sahara. https://wiki.openstack.org/wiki/Sahara
15. Overlay Filesystem. https://goo.gl/y0H76w
16. Puppet. http://puppetlabs.com/
17. Rackspace SLAs. https://www.rackspace.com/information/legal/cloud/sla
18. Vagrant. https://www.vagrantup.com/
19. VMware vCloud Automation Center Documentation Center. http://goo.gl/YkKNic
20. Jennings, B., Stadler, R.: Resource management in clouds: survey and research challenges. J. Netw. Syst. Manag. **23**(3), 567–619 (2015)
21. Juve, G., Deelman, E.: Automating application deployment in infrastructure clouds. In: 2011 IEEE Third International Conference on Cloud Computing Technology and Science (CloudCom), pp. 658–665. IEEE (2011)
22. Katsuno, Y., Takahashi, H.: An automated parallel approach for rapid deployment of composite application servers. In: 2015 IEEE International Conference on Cloud Engineering (IC2E), pp. 126–134. IEEE (2015)
23. Liu, C., Mao, Y., Van der Merwe, J., Fernandez, M.: Cloud resource orchestration: a data-centric approach. In: Proceedings of the biennial Conference on Innovative Data Systems Research (CIDR). pp. 1–8 (2011)
24. Lu, Q., Zhu, L., Xu, X., Bass, L., Li, S., Zhang, W., Wang, N.: Mechanisms and architectures for tail-tolerant system operations in cloud. In: 6th USENIX Workshop on Hot Topics in Cloud Computing (HotCloud 14) (2014)
25. Mell, P., Grance, T.: The NIST Definition of Cloud Computing (2011)
26. Potharaju, R., Jain, N.: When the network crumbles: an empirical study of cloud network failures and their impact on services. In: Proceedings of the 4th Annual Symposium on Cloud Computing, p. 15. ACM (2013)

27. Rodeh, O., Bacik, J., Mason, C.: Btrfs: the linux b-tree filesystem. ACM Trans. Storage (TOS) **9**(3), 9 (2013)
28. Tsoumakos, D., Konstantinou, I., Boumpouka, C., Sioutas, S., Koziris, N.: Automated, elastic resource provisioning for nosql clusters using tiramola. In: 2013 13th IEEE/ACM International Symposium on Cluster, Cloud and Grid Computing (CCGrid), pp. 34–41. IEEE (2013)

Fair, Fast and Frugal Large-Scale Matchmaking for VM Placement

Nikolaos Korasidis[1]([⊠]), Ioannis Giannakopoulos[1], Katerina Doka[1],
Dimitrios Tsoumakos[2], and Nectarios Koziris[1]

[1] Computing Systems Laboratory,
National Technical University of Athens, Athens, Greece
{nkoras,ggian,katerina,nkoziris}@cslab.ece.ntua.gr
[2] Ionian University, Corfu, Greece
dtsouma@ionio.gr

Abstract. VM placement, be it in public or private clouds, has a decisive impact on the provider's interest and the customer's needs alike, both of which may vary over time and circumstances. However, current resource management practices are either statically bound to specific policies or unilaterally favor the needs of Cloud operators. In this paper we argue for a flexible and democratic mechanism to map virtual to physical resources, trying to balance satisfaction on both sides of the involved stakeholders. To that end, VM placement is expressed as an Equitable Stable Matching Problem (ESMP), where each party's policy is translated to a preference list. A practical approximation for this NP-hard problem, modified accordingly to ensure efficiency and scalability, is applied to provide equitable matchings within a reasonable time frame. Our experimental evaluation shows that, requiring no more memory than what a high-end desktop PC provides and knowing no more than the top 20% of the agent's preference lists, our solution can efficiently resolve more than 90% of large-scale ESMP instances within $N\sqrt{N}$ rounds of matchmaking.

1 Introduction

VM placement is a fundamental problem addressed by current cloud providers [1–3]. The policy through which the VMs are placed into the physical hosts tremendously affects the data center's utilization, energy consumption and operational cost [1] while, at the same time, it greatly influences the VM performance and, hence, the cloud customer satisfaction. Albeit initially static and determined solely by the capacity and the utilization of the physical hosts, the VM placement schemes are becoming more sophisticated and democratic, taking into consideration the client's needs as well. Indeed, policies that increase both the data center's efficiency and the VMs' performance can prove a boon for providers and clients alike, keeping both sides content. For example, a user would benefit from collocating two VMs that require a low-latency network connection whereas the provider would also benefit from packing those VMs into a single host to save

© Springer International Publishing AG 2017
T. Sellis and K. Oikonomou (Eds.): ALGOCLOUD 2016, LNCS 10230, pp. 131–145, 2017.
DOI: 10.1007/978-3-319-57045-7_8

network bandwidth. In the general case, though, the interests of the two parties may also oppose each other, a fact that complicates decision-making.

Some works in the field examine the policy-based VM placement from a game-theoretic viewpoint, formulating the problem as a multi-agent game in which each agent tries to maximize her profit [9,15]. The provided solution is a personalized pricing scheme, which mainly targets fairness among users themselves, rather than users and providers. Other approaches view the placement problem as a matching problem [11,13,14]. Instead of attempting to optimize a multi-dimensional utility function that encapsulates the objectives of both providers and users, each stakeholder retains a *preference list* that expresses her policy. The problem is then transformed into the Stable Marriage Problem (SMP) [7], where the objective is to identify a stable matching between VMs/VM slots.

Stability is a key concept in this formulation as it guarantees that there exists no pair of a VM (representing the cloud users) and a VM slot (representing the data center) that prefer each other more than their current match. Such a matching is proven to exist in any problem instance and it can be found by the Gale-Shapley algorithm [7]. Although the Gale-Shapley algorithm guarantees stability, it fails to ensure that the two sides are treated equally. On the contrary, due to the strict proposer-acceptor scheme it relies on, it inherently favors one side of the involved negotiators. Besides, finding an optimally fair stable matching is proven to be NP-hard [10]. Furthermore, the execution of the Gale-Shapley algorithm requires $\mathcal{O}(N^2)$ memory, N being the number of VMs, rendering it unsuitable for resource management in modern data centers that host hundreds of thousands of VMs.

To overcome the limitations of unfairness and quadratic memory growth, we propose a novel VM placement approach that seeks for a fair solution based on incomplete preference lists. Specifically, we generate preference lists based on the provider and user policies and apply the heuristic from our previous work in [8] to create a fair matching between the VMs and the available VM slots. To avoid quadratic memory expansion and scale to thousands of VMs, we reduce the preference lists to the top-K most preferable positions, rejecting any proposal originating from an agent lower on the list than the Kth one. Since this optimization modifies the problem's properties, we evaluate its impact on the algorithm execution and showcase that the usage of incomplete lists is suitable for most problem instances since equivalent matchings can be produced using only a portion of the original lists. Furthermore, through our extensive evaluation we study the parameters that affect the performance of our VM placement methodology and demonstrate that we can provide fair solutions even for problem instances up to 20 k VMs within a few seconds, while using only 20% of the memory needed in the casual case without compromising neither the correctness nor the quality of the algorithm. Our concrete contributions can be, thus, summarized as follows:

- We formulate the policy-based VM placement problem as the sex-equal SMP and utilize an approximation algorithm [8] to identify a fair solution.

- We introduce a variation of the original approximation algorithm that relies on incomplete preference lists to be able to accommodate larger problem instances.
- We provide a thorough experimental study that evaluates our proposed method under various situations arising in modern data centers and demonstrates that our approach is particularly suitable for large-scale datasets where thousands of VMs must be scheduled for deployment.

2   Related Work

VM placement is a vividly researched problem in the cloud computing area. The suggested approaches can be categorized in two distinct classes: (a) the ones that aim to resolve the problem favoring exclusively the provider and (b) the ones that seek for a compromise between the provider's interest and the user-defined placement policies.

The first category includes approaches that optimize data center indicators such as utilization, operational cost, energy consumption, etc., while honoring the Service Level Agreements (SLAs). [2] focuses on the initial VM placement when a stream of deployment requests arrive to a data center. [1] proposes an energy efficient VM placement scheme that utilizes migration to relocate VMs so as to decrease the operational cost without jeopardizing the Quality of Service (QoS). In [4], a classification methodology is proposed with the objective to maximize the data center's efficiency, while the VM performance is kept at maximum. Similarly, in [12] the authors propose a discrete knapsack formulation, where the objective is to maximize a global utility function that corresponds to the operational cost with the constraint of respecting the SLAs. The authors utilize a Constraint Programming approach to tackle the problem. None of the above methods take into consideration user-defined scheduling policies.

Regarding the second category, we encounter two distinct problem formulations. The first, involves a game-theoretic approach, in which each customer is viewed as an agent that attempts to maximize utility. The objective of such formulations is to achieve an equilibrium in the game, so that the players are treated equally. For example, in [9], an approach is that attempts to provide personalized prices to each customer is presented. The idea is that each customer defines their workload and the provider needs to specify a price that is fair among the customers and is also beneficial for both the customer and the provider. Similarly, Xu and Yu [15] present an approach that targets to provide fairness among the clients and also increase the cluster utilization.

Alternatively, the policy-based VM placement is formulated as an SMP, in which VMs are matched to the available hosts. This formulation entails the extraction of a preference list for each entity (i.e., VM and VM slot) that reflects the policies of the involved parties. [14] proposes a system that accepts user policies and executes the Gale-Shapley algorithm to find a mapping between VMs and physical resources while [11] uses the same algorithm to improve energy consumption. The resulting match in both cases, however, is proposer (i.e., Cloud provider)-optimal.

In our work, we utilize ESMA [8] to provide fair and stable solutions and propose an enhancement thereof to deal with the quadratic expansion of memory requirements as the instance size increases. Equality in SMP is also considered in [13], where out of a number of different matchings constructed, the algorithm selects the one that maximizes fairness between the opposing groups. Contrarily, our approach produces a single fair solution, with performance comparable to the non-equal Gale-Shapley algorithm. This renders our work more suitable for real-time execution as new deployment requests arrive at the data center.

3   Preliminaries

An instance I of the *stable marriage problem* (SMP) consists of n men and n women, where each person has a *preference list* that strictly orders all members of the opposite gender. If a man m prefers w_1 to w_2, we write $w_1 \succ_m w_2$; a similar notation is applied to women's preferences. A *perfect matching M* on I is a set of disjoint man-woman pairs on I. When a man m and a woman w are matched to each other in M, we write $M(m) = w$ and $M(w) = m$. A man m and a woman w are said to form a *blocking pair* for M (or to *block M*) when: (i) $M(m) \neq w$; (ii) $w \succ_m M(m)$; and (iii) $m \succ_w M(w)$. A matching M is *unstable* if a blocking pair exists for M, and *stable* otherwise. The SMP calls for finding a stable matching M.

The standard algorithm for solving the SMP, proposed by Gale and Shapley [7], goes through a series of iterations. At each iteration men propose and women accept or reject proposals. The algorithm is guaranteed to terminate successfully after a quadratic number of steps, providing a perfect matching which, in effect, cater to the satisfaction of the proposers. Since many different stable matchings may exist for any problem instance, it is reasonable to aim for a matching that is not only stable, but also good by some quality metric.

3.1   Equitable Stable Marriage Problem

Past research has defined three quality criteria. Let $pr_m(w)$ (respectively, $pr_w(m)$) denote the position of woman w in man m's preference list (respectively, of m in w's list). The *regret cost $r(M)$* of a stable matching M is:

$$r(M) = \max_{(m,w)\in M} \max\{pr_m(w), pr_w(m)\} \tag{1}$$

On the other hand, the *egalitarian cost $c(M)$* is:

$$c(M) = \sum_{(m,w)\in M} pr_m(w) + \sum_{(m,w)\in M} pr_w(m) \tag{2}$$

Finally, the *sex equality cost* is defined as:

$$d(M) = \left| \sum_{(m,w)\in M} pr_m(w) - \sum_{(m,w)\in M} pr_w(m) \right| \tag{3}$$

A stable matching that optimizes the sex equality cost satisfies a notion of equity between the two sides of the involved stakeholders. Finding such a solution to the so-called *equitable stable marriage problem* (ESMP) is NP-hard [10], thus all proposed solutions use heuristics that either produce different stable matchings and seek for the fairest at the cost of increased execution time, or attempt to construct an equal approach by allowing proposers and acceptors to interchange their roles. The execution time penalty of the former category is prohibitive for our use case, since we need to be able to produce matchings in an online fashion. The Equitable Stable Matching Algorithm (ESMA) [8] is an algorithm of the latter category, that allows the opposite group members to both propose and accept in different algorithm steps. It is experimentally proven that ESMA, unlike its main competitor [6], (a) terminates on all tested large problem instances, (b) generates equitable matchings of high quality and (c) has an execution time similar to the Gale-Shapley algorithm, outperforming other similar solutions.

ESMA utilizes a simple non-periodic function in order to assign the proposer group. Each proposer proposes to their next preference and the acceptors evaluate their proposals and accept their most preferred choice rejecting the others, as in the classic Gale-Shapley algorithm. When the acceptors become proposers, they also start proposing to their most preferred choices only if the agents they propose to are most preferable than their current fiancé. Each agent retains two indices into their preference list: one that indicates their current fiancé and one that indicates their current most desired agent. These indices change positions while new proposals are issued and new marriages are established. The algorithm terminates when all agents of both groups have concluded that they are married to their most preferred choice that did not reject them, i.e., the two aforementioned indices point to the same agent.

As discussed in [8], this alternating proposing scheme may introduce repeating proposal patterns since two agents from the opposite groups may establish more than one engagements between them during the algorithm's execution, something which is not allowed in the Gale-Shapley algorithm due to the *monotonic* preference list traversal of the agents, i.e., the proposers start from the beginning of their lists and degrade as the algorithm evolves and the opposite stands from the acceptors. Since such a repeating pattern may compromise the algorithm's termination by creating endless proposal loops, the non-monotonic function utilized for assigning the proposer group seems to overcome this challenge and leads to termination for all the tested problem instances. In our work, we have not encountered any problem instance that lead ESMA to non-termination due to repeating cycles. This feature, in combination with the low per-step complexity ($\mathcal{O}(n)$) and the short execution times, establish ESMA as the basis for the VM placement scheme we propose at this work. Its application for VM placement is straightforward: We assume that one side consists of the clients' VMs to be deployed and the other side consists of the available VM slots. The final matching determines the slot that each VM will occupy.

3.2   Modeling Preferences

The extraction of the preference lists from the policies defined by the user and the cloud provider is a tedious task, tightly coupled with the policy model. Various transformation schemes exist, depending on the nature of the policy. For example, [14] assumes that the users define their policies in a rule-based format, easily transformable to a serialized preference list for the available VM hosts. Specifically, the user generates *policy groups* in which the preference for a host is expressed in the form: $1/\text{CPU}_{freq}$ (the shorter the number the higher the preference), meaning that the higher CPU frequency hosts should be preferred. Composition of such simple rule-based policies would create a serialized preference list for each user and the provider, that also creates policies on a similar manner, based on data center metrics such as resource utilization, energy consumption, etc. On the contrary, in [9], a different policy scheme is adopted, where the user defines a graph that demonstrates the type of needed resources (e.g., cores, memory, etc.) and the timeline of their utilization. Obviously, such a policy description would entail a totally different preference list extraction algorithm. In this work, we mainly focus on what happens *after* the policy extraction. Since different schemes could be adopted in different use cases, we assume a simple policy mechanism like the one introduced in [14] and study various aspects of our approach when policies produce preference lists of different distributions and properties.

4   Dataset Generator

To test our approach, we made use of synthetic data, i.e., preference lists generated using *Reservoir Sampling*. This is easy to implement and properly models the case where all agents are of the roughly same quality/fitness. However, in real life applications it is often the case that some portion of the resources available are globally more preferable to others. This variation in quality can be discreet (e.g., high quality agents VS low quality agents) or continuous. To incorporate this possibility, we assume that a weighting function exists that maps each agent to some *global bias*: agents with high global bias are more likely to end up in (the top places of) the preference lists of their prospective suitors. Thus, we implemented instance generation via *Weighted Reservoir Sampling*, following the A-Res interpretation [5]: If the sums of all weights/biases is B, an agent with bias b_i has probability b_i/B to appear in their suitors top preferences. Despite the biases, each agent's list is generated independently; randomness is provided via an agent-specific 32-bit *seed* which is derived from a global seed. Instance generation is entirely portable and reproducible, as it is based on PRNGs available in C++11.

Bias modeling happens through allocating different weights to different agents, and feeding those weights to WRS. We studied three different weight distributions, namely *uniform distribution*, *contested distribution* and *position-inverse distribution*.

In the uniform distribution (UNI) each agent is assigned the same, constant bias. This produces preference lists that are selected uniformly from all $K-$ permutations.

In the contested distribution (CONT), the agents are partitioned into two sets; one set comprises of higher-quality suitors that are globally preferable, whereas the other set is medium-quality suitors. Agents in the high-quality set are assigned the same, constant weight that is 10 times larger than the weight assigned to agents in the lower-quality set. A parameter p controls the fraction of the agents that are considered high-quality. In our experiments we used $p = 0.1$.

Finally, in the position-inverse distribution (INV), each agent is assigned a weight that is a falling function of its id in the set, i.e. $i < j \implies b_{a_i} > b_{a_j}$. In our experiment, we used the function $f_a(i) = (i + 1)^{-a}$ to map integer ids to positive weights, for $a = 1.0$.

The reason for choosing these weight distributions is to model circumstances arising in the VM placement problem. Two parameters are of particular interest: the variation of a data center's equipment as time passes, and the variability in client's performance demands.

A new data center usually contains machines of about the same overall quality, which corresponds to UNI. Doing a minor hardware upgrade or expansion results in a small number of new machines becoming more powerful than the rest, a situation that corresponds to CONT. The cumulative effect of many such upgrades is a collection of machines with smoothly varying performance characteristics, corresponding to INV. The 2 latter ways of bias modeling result in preference lists that are not uniformly random $K-$ permutations: we expect the top fraction of those lists to be somewhat similar, creating a tougher problem instance.

Similarly, clients of a data center may be diversified in the quality of service they demand. The data center may provide only one level of QoS (modeled as UNI). Alternatively, it may provide two levels, an economical plan and a premium one (modeled as CONT). Finally, it may have some more refined service program, where the client pays more according to a smoothly varying parameter ("nines of uptime"), which we model with INV.

Out of the nine possible situations of biases, we have chosen those with the same bias on each side. This is done for simplicity but is not detached from the real world. A small scale data center would probably serve customers without special needs. On the contrary, any large data center serving a diversified clientele is probably old enough to have amassed equipment of varied performance.

5   Intuiting ESMP with Trimmed Lists

According to our formulation, achieving good VM placement reduces to solving large instances of ESMP. Unfortunately, the data set scales quadratically with N: an ESMP instance of 20 k agents per set requires more than 6 GB of space. Solving larger instances requires resources usually available only in very high-end machines. Thus, our primary goal is to drastically reduce ESMA's memory footprint.

5.1   An Interesting Metric

While running ESMA over randomly generated instances with at least 5000 agents, we are surprised to obtain solutions of low regret cost, despite never explicitly trying to optimize this cost. More importantly, the regret cost as a fraction of N *falls* as N grows larger, or, stated in plain words, as the pool of suitors expands, ESMA becomes increasingly effective at matching all agents to their top preferences. This low regret cost could just be a (welcome) artifact of the final matching, however, the statistics from our runs reveal a more systematic cause. It turns out that in any successful ESMA run all agents propose solely to their top preferences *throughout the matchmaking.* If we track the "proposer regret cost" over all rounds, we see that its maximum is only slightly larger than the final regret cost (the latter taken over all agents). So, although an agent acting as an acceptor may get proposed by a unfit suitor, no agent acting as a proposer does likewise. The following box-n-whisker plots are telltale (Fig. 1).

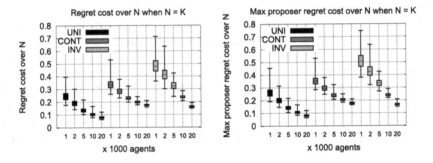

Fig. 1. Regret cost and max proposer regret cost for complete preference list

While proposers behave in a way that guarantees a low regret cost, acceptors may not necessarily do so: they may accept a low-ranking suitor as a partner. However, such an arrangement is not satisfying to them: For an agent to be satisfied according to ESMA, they must be matched to a partner that is within the fitness range they tolerate, which in turn is specified by the suitors they have been rejected from. As we noted already, an ESMA agent only proposes to (and hence is rejected from) suitors high in their preference list. Thus, for an agent to become satisfied when *accepting* a proposal, the proposer must be highly-preferred. An unsatisfactory match cannot be permanent. Therefore, when all agents are satisfied, it is because *all* are matched to a highly preferred partner.

5.2   Our Approach

Intuitively, since no agent proposes to or forms a stable match with a lowly-preferred suitor, a fraction of the preference lists is effectively irrelevant to the solution: the ids of lowly-preferred suitors as well as the entries in the inverse map that encodes the fitness of said suitors. We set out to investigate if it is possible to solve ESMP instances successfully while restraining ourselves to using only the topmost part of the preference lists, i.e. the part that encodes highly-preferred suitors. Our results, outlined in a following section, reveal that this is indeed the case; the majority of large instances can be successfully solved with at most 20% of the full data set, without sacrificing the quality of the solution.

5.3 Proof of Stability

We can prove that whenever the algorithm terminates properly, the resulting matching is stable despite the incomplete lists. Consider an instance with N agents per set, each of whom has trimmed lists of length K. If $K = N$, we can rule out the existence of a blocking pair by invoking the stability proof cited in [8]: once all agents are satisfied, the resulting matching is stable and no agent can expect to be accepted by a more preferred suitor than their partner. In the case where $K < N$, we can show that no extra blocking pairs can form once the full lists are revealed, by reductio ad absurdum. Suppose such a pair (m_i, w_j) materialises only when they get to know their full lists, and that m_i prefers w_j to its current partner w_{m_i}. For m_i to sport such a preference only after the full lists are revealed, w_j must be among the trimmed preferences of m_i. But w_{m_i} must rank lower than w_j in his preference list, thus w_{m_i} also lies on the trimmed part, therefore m_i could never have engaged with her, a contradiction. Thus, no new blocking pairs form. Since all agents are satisfied, there were no blocking pairs initially. Combined, these two arguments prove that no blocking pairs exist at all; solving ESMP with trimmed lists produces stable matching given termination.

6 Experimental Evaluation

Driven by the previous plots of low regret cost, we set out to experimentally investigate how much data can we ignore and still match agents successfully. Our testsuite consists of ESMP instances that vary according to three main parameters: number of agents per set N, fraction of preference list available K/N and bias modelling. We test all combinations of the following with 500 different seeds each:

- $N \in \{1000, 2000, 5000, 10000, 20000\}$
- $K/N \in \{0.1, 0.2, 0.5, 1.0\}$
- uniform bias, contested bias for $p = 0.1$, position inverse bias for $a = 1.0$

6.1 Ensuring Correctness

We declare an execution to be SUCCESSFUL if it terminates properly with a matching within $N\sqrt{N}$ rounds of matchmaking, otherwise it is declared a FAILURE. The reason for choosing this particular round limit will be addressed in a later section, along with an investigation of failed runs. In the following graphs, we plot the percentage of successful executions for each combination of parameters.

We observe that for any given fraction and bias tested, the percentage of successful executions increases with N or remains practically the same. This means that our approach is better suited for large, memory-demanding ESMP instances than for small ones. A most interesting finding is that the success rate initially increases as the lists get smaller but becomes zero after a threshold is reached, e.g., for 1 k–5 k agents in the Contested distribution case for $K/N = 20\%$. There appears to be a sweet spot for the list size, which depends on the problem size and weights used. We interpret this result as follows: As discussed earlier, no acceptor is stably matched to a suitor they prefer little. Reducing the list size outright prevents an acceptor from entertaining such a temporary match and forces the proposer to keep searching instead of idling. This reduces the number of rounds required until a globally stable match is reached, hence more executions terminate correctly within the given round limit. However, if the lists are trimmed too much, there are not enough options to accommodate everyone and hence the execution fails, as is expected in SMP with Incomplete Lists (Fig. 2).

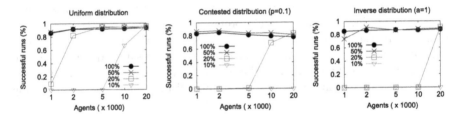

Fig. 2. Regret cost and max proposer regret cost for complete preference list

Other experiments, omitted due to space constraints, revealed that the "sweet spot" effect diminishes or even reverses completely if we allow for a very large round limit, e.g. in the order of $\mathcal{O}(N^2)$ In particular, all full-list instances terminate if given enough rounds, while some trimmed-lists instances seem to run forever. This phenomenon is interesting from a theoretical as well as a practical viewpoint.

As expected, uniform bias produces the easiest instances, position-inverse bias comes up second and contested bias is overall the hardest one. At $N = 20000$ and using only 20% of the full lists, we can solve more than 90% of instances of uniform and position-inverse bias, and more than 80% of instances produced via contested bias, using about 610 MB of RAM and at most about 2.8 million rounds. This demonstrates the power of our approach. In terms of execution, our approach needed approximately 15 s for the $N = 20$ k, $K/N = 100\%$ case whereas for the same number of agents and smaller preference list (e.g., $K/N = 10\%$) the algorithm only took 6.5 s. Both experiments were executed on a Linux machine using a 24 threaded Xeon X5650 CPU at 2.67 GHz. The observed acceleration is attributed to the fact that shorter preference lists lead to faster access to the data structures which are allocated to support them. For more information, the reader can consult the Appendix A.

6.2 Maintaining Quality

Increased performance often comes at the price of quality. One would expect that the reduced amount of information causes our algorithm to select a worse matching. Fortunately, this effect appears to be quite mild, as detailed in the following plots. The regret and egalitarian costs remain virtually constant. The average sex-equal cost is negatively impacted, especially if N is small and we are not using uniform weights, but the difference is quite small. The variation of seq-equal cost increases the more we trim the lists. Again, the quality degradation is more intense for smaller datasets, something that indicates that our approach is more suitable for massive datasets, rather than smaller instances of ESMP which are easily solved by commodity machines (Fig. 3).

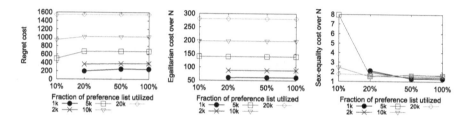

Fig. 3. Matching costs for different fractions of preference lists

6.3 Failing Gracefully

As evidenced in the above graphs, no execution terminates successfully if preference lists are trimmed too much. This reduces the usefulness of our approach, as we cannot always predict the proper K for an instance a priori. We rectify this defect by showing that ESMA fails gracefully: it succeeds in matching the vast majority of agents to satisfying partners, even when a total matching is impossible (Fig. 4). This happens even under extreme constraints: we run 500 ESMP instances with $N = 20000$ agents and lists trimmed at $K = 1000$, and recorded how many rounds passed until a sufficiently large fraction p of proposers are satisfied with their current partners, for $p \in \{90\%, 95\%, 99\%, 99.5\%, 99.9\%, 99.95\%, 99.99\%\}$. Our results appear below (Table 1).

While none of the executions terminated successfully, most were able to come very close to complete stable matches in a surprisingly low number of rounds: 99.9% of proposers had been satisfied with their matches after about 30000

Table 1. Matched agents vs number of runs

Percentile matched	90	95	99	99.5	99.9	99.95	99.99
Gracefully failed runs	500	500	500	500	420	2	0

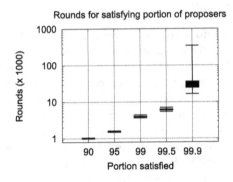

Fig. 4. Matched agents vs number of runs

rounds, that is, within 1% of the allowed round limit. Since our instances used extremely small preference lists, it is reasonable to assume that a similar amount of acceptors were also satisfied by the same number of rounds. Thus, after only few rounds of matchmaking, we can have a partial plan that places most VMs in appropriate slots and keeps most customers satisfied. The few resources that were not properly matched can be dealt with in an ad hoc way.

It should now be obvious that the arbitrary $N\sqrt{N}$ round limit was chosen out of pragmatic concerns. For large testcases, $\mathcal{O}(N^2)$ rounds of matchmaking (required by the GS algorithm to produce a complete stable matching in the worst case) is a prohibitively large amount given the setting of our problem. On the other hand, $\mathcal{O}(N)$ rounds seem to suffice for an almost complete partial match. We chose an intermediate value, so that our results can be used to generate useful guidelines. Depending on the specific workload and balancing the need for speed and completeness, a provider can choose their own round limit.

7   Conclusions

In this paper we revisited the problem of policy-based VM placement in a modern data center. In the equitable methodology we propose, an approximation algorithm was utilized in order to find a stable and fair matching between the VMs and the available VM slots, so as to honor the policies dictated by the opposite parties. Moreover, we developed an optimization of the original algorithm, in which we trimmed the preference lists of both groups to enable its scaling to hundreds of thousands of VMs, a typical case for modern data centers. Through an extensive experimental evaluation we showcase that our approach is able to find fair matchings for up to 20 k VMs, which is an order of magnitude larger than other competitive approaches. Simultaneously, we demonstrate that trimming the preference lists is particularly effective for large cases, since both the correctness and the quality of ESMA's solutions, measured in terms of equality and global happiness, are maintained. As a future work, we seek confirmation of the results presented at this work with real-world workloads gathered in public data centers. Furthermore, we envision to port our methodology into streaming deployment requests so as to serve the needs of rapidly changing demands.

Appendix

A   Implementation details

Our implementation is a rewrite of the one presented in [8], with a few changes and additions to suit our purpose. The source code can be found on GitHub: https://github.com/Renelvon/pesma-limit.

A.1   Languages and Frameworks

The original Java implementation was ported to C++11, as the later allows much greater precision in handling memory resources. We took care to make as much of the new code parallelizable and used the OpenMP framework to leverage any multicore power during instance generation, loading and solving.

A.2   Proposer selection

We propose a minor modification to the proposer-selection heuristic used in [8]: whenever all agents of one set are satisfied after some round, the other set is granted the proposer role in the next round. This ensures there will be some proposing action in all rounds save for the last one. It prevents idle rounds in large instances, and is especially impactful once only a few agents remain unsatisfied. Despite the tangible benefits, it is unclear whether this optimization may force the proposer-selection algorithm to fall in a periodic cycle, though none of our experiments indicate so.

A.3   Deterministic generation and subsetting

A problem instance generated by our framework can be reliably reproduced given the quintuple (N, n, K, b, s, h), where N is the number of agents per set, n is the size of the reservoir used, K is the length of the preference list, b is a function mapping agents to *positive* weights (floating-point values), s is an initial seed (64-bit integer) and h is a hash function. We use the 64-bit Mersenne Twister (MT64) engine available in the Standard Library of C++11, a fast PRNG that guarantees long periods.

We generate a preference list as follows. First, we hash the global seed s together with the agent id to create an agent-specific seed s', with which we setup a MT64. We feed the engine and the function b together with N and n to WRS and obtain a sample of suitors of length n, where $n \leq N$. Each sampled suitor is paired with the weight which put them in the reservoir. As a last step, we extract the heaviest K suitors (where $K \leq n$) in the sample and place their ids in the preference list ordered according to their weight. Each agent's list can be generated independently; our OpenMP implementation takes advantage of this.

Consider two instances occurring from the same (N, n, b, s, h) but differring in the cut-off point: $K_1 < K_2 \leq n$. The sampling phase is independent from K and so will produce the same sample of length n for any two agents with the same id. Since the suitors are placed in the preference list according to the order of their weights, the list of an agent in the K_1 instance will coincide with the first K_1 elements of the corresponding agent's (longer) list in the K_2 instance. This effect enables us to generate instances that are *proper subsets* of other instances. We can thus study with great precision the effects of limiting the preference list at any desired fraction.

A.4   Compact representation of inverse map

The bulk of each agent's dataset consists of two structures: the preference list, which contains the suitors in order of preference, and the inverse map, which maps a suitor's id to their index in the agent's preference list. When entire preference lists are used, the inverse map can be represented as a plain array. However, once the preference lists are trimmed, the array becomes sparse and it is wise to represent it as a static dictionary in order to save space. We explored several concrete implementations built around `stl::unordered_map` and `boost::flat_map`. The former one is very fast but the latter one achieves a much better balance between performance and memory consumption. For our experiments, we reimplemented `boost::flat_map` from scratch to reduce unnecessary overheads and augmented it with a summarizer cache that enables faster searches.

References

1. Beloglazov, A., Buyya, R.: Energy efficient resource management in virtualized cloud data centers. In: Proceedings of the 2010 10th IEEE/ACM International Conference on Cluster, Cloud and Grid Computing, pp. 826–831. IEEE Computer Society (2010)
2. Calcavecchia, N.M., Biran, O., Hadad, E., Moatti, Y.: VM placement strategies for cloud scenarios. In: 2012 IEEE 5th International Conference on Cloud Computing (CLOUD), pp. 852–859. IEEE (2012)
3. Chaisiri, S., Lee, B.-S., Niyato, D.: Optimal virtual machine placement across multiple cloud providers. In: Services Computing Conference (APSCC 2009), IEEE Asia-Pacific, pp. 103–110. IEEE (2009)
4. Delimitrou, C., Kozyrakis, C.: Quasar: resource-efficient and QoS-aware cluster management. In: ACM SIGPLAN Notices, vol. 49, pp. 127–144. ACM (2014)
5. Efraimidis, P.S., Spirakis, P.G.: Weighted random sampling with a reservoir. Inf. Process. Lett. **97**(5), 181–185 (2006)

6. Everaere, P., Morge, M., Picard, G.: Minimal concession strategy for reaching fair, optimal and stable marriages. In: Proceedings of the 2013 International Conference on Autonomous Agents and Multi-agent Systems, pp. 1319–1320. International Foundation for Autonomous Agents and Multiagent Systems (2013)
7. Gale, D., Shapley, L.S.: College admissions and the stability of marriage. Am. Math. Monthly **69**(1), 9–15 (1962)
8. Giannakopoulos, I., Karras, P., Tsoumakos, D., Doka, K., Koziris, N.: An equitable solution to the stable marriage problem. In: 2015 IEEE 27th International Conference on Tools with Artificial Intelligence (ICTAI), pp. 989–996. IEEE (2015)
9. Ishakian, V., Sweha, R., Bestavros, A., Appavoo, J.: Dynamic pricing for efficient workload colocation (2011)
10. Kato, A.: Complexity of the sex-equal stable marriage problem. Jpn. J. Ind. Appl. Math. **10**(1), 1–19 (1993)
11. Kella, A., Belalem, G.: VM live migration algorithm based on stable matching model to improve energy consumption and quality of service. In: CLOSER, pp. 118–128 (2014)
12. Van, H.N., Tran, F.D., Menaud, J.-M.: SLA-aware virtual resource management for cloud infrastructures. In: Ninth IEEE International Conference on Computer and Information Technology (CIT 2009), vol. 1, pp. 357–362. IEEE (2009)
13. Xu, H., Li, B.: Egalitarian stable matching for VM migration in cloud computing. In: 2011 IEEE Conference on Computer Communications Workshops (INFOCOM WKSHPS), pp. 631–636. IEEE (2011)
14. Xu, H., Li, B.: Anchor: a versatile and efficient framework for resource management in the cloud. IEEE Trans. Parallel Distrib. Syst. **24**(6), 1066–1076 (2013)
15. Xu, X., Yu, H.: A game theory approach to fair and efficient resource allocation in cloud computing. Math. Probl. Eng. **2014**, 1–14 (2014)

Survey of Machine Learning Algorithms on Spark Over DHT-based Structures

Spyros Sioutas[1], Phivos Mylonas[1], Alexandros Panaretos[1],
Panagiotis Gerolymatos[1], Dimitrios Vogiatzis[1], Eleftherios Karavaras[1],
Thomas Spitieris[1], and Andreas Kanavos[2]([✉])

[1] Department of Informatics, Ionian University, Corfu, Greece
{sioutas,fmylonas,alex,c12gero,p12vogi,p12kara,p12spit}@ionio.gr
[2] Computer Engineering and Informatics Department,
University of Patras, Patras, Greece
kanavos@ceid.upatras.gr

Abstract. Several solutions have been proposed over the past few years on data storage, data management as well as data retrieval systems. These solutions can process massive amount of data stored in relational or distributed database management systems. In addition, decision making analytics and predictive computational statistics are some of the most common and well studied fields in computer science. In this paper, we demonstrate the implementation of machine learning algorithms over an open-source distributed database management system that can run in parallel on a cluster. In order to accomplish that, a system architecture scheme (e.g. Apache Spark) over Apache Cassandra is proposed. This paper also presents a survey of the most common machine learning algorithms and the results of the experiments performed over a Point-Of-Sales (POS) data set.

Keywords: Apache spark · Apache cassandra · Big data · Data analytics · DHT-based Structures · Machine learning

1 Introduction

The rapid development of technology provides high quality network services. A large percentage of users utilize the Internet for information in various fields. Companies try to take advantage of this situation by creating systems that store information about users who entered their site in order to provide them with personalized promotions. Moreover, companies concentrate on their desired information and personal transactions and on the other hand, businesses provide their customers with cards so they can record every buying detail. This procedure has led to a huge amount of data and search methods for data processing. As a previous work on data processing emerging in reviews, one can consider the setup presented in [3]. Furthermore, in [8], authors infer details on the love bond between users and a brand name being considered as a dynamic ever evolving

T. Sellis and K. Oikonomou (Eds.): ALGOCLOUD 2016, LNCS 10230, pp. 146–156, 2017.
DOI: 10.1007/978-3-319-57045-7_9

relationship. The aim of this work is to find those users that are engaged and rank their emotional terms accordingly.

Historically, several analysts have been involved in the collection and processing of data. In modern times, the data volume is so huge that it requires the use of specific methods so as to enable analysts to export correct conclusions. Due to the increased volume for automatic data analysis, methods use complex tools; along with the help of modern technologies, data collection can be now considered as a simple process. Analyzing a dataset is a key aspect to understanding how customers think and behave during each specific period of the year. There are many classification and clustering methods which can be successfully used by analysts to aid them broach in consumers' mind. More specifically, supervised machine learning techniques are utilized in the present manuscript in the specific field of Point-Of-Sales (POS).

In addition, frameworks like Hadoop, Apache Spark, Apache Storm as well as distributed data storages like HDFS and HBase are becoming popular, as they are engineered in a way that makes the process of very large amounts of data almost effortless. Such systems are gaining much attention and consecutively libraries (like Apache Spark's MLlib), which make the use of Machine Learning techniques possible in the cloud, are introduced.

In an effort to experiment with some of the most influential algorithms that have been widely used in the machine learning community, an implementation on Apache Spark over an open-source DHT-based database management system Apache Cassandra is presented. We chose to implement Naive Bayes, SVM, K-Means, Expectation Maximization, Association Rules and Parallel FP-Growth which are some of the top algorithms in Data Mining [16,17].

The remainder of the paper is structured as follows: Sect. 2 presents the open-source interface Apache Spark as well as the Apache Cassandra distributed database management system. Section 3 presents the Machine Learning Algorithms which were utilized in our proposed system, while Sect. 4 presents the Point-Of-Sales data that we use for our experiments. Moreover, Sect. 5 presents the evaluation experiments conducted and the results gathered. Finally, Sect. 6 presents conclusions and draws directions for future work.

2   Apache Spark and Apache Cassandra

Apache Spark[1] is a widely known open-source Application Programming Interface (API) for cluster computing which provides programmers with the ability to rapidly develop applications using languages like Scala (which is the native language that Spark is written), Java, Python or R in order to process very large data sets. Spark is based on an in-memory programming model thus it overcomes the multiple and time expensive I/O problems that Hadoop Map-Reduce introduces. It also provides scalability and fault tolerance as main characteristics and functionalities.

---

[1] http://spark.apache.org/.

Spark architecture relies on a data structure called resilient distributed data set (RDD) which represents a collection of objects partitioned across a set of machines. RDDs are cached in memory and can be used in multiple parallel MapReduce-like operations [18, 19]. Spark consists of a core system that contains features like task scheduling, memory management and fault recovery techniques and a set of abstractive libraries that rest on top. These libraries, as shown in Fig. 1, provide a rich set of higher-level tools including Spark SQL for SQL and structured data processing, MLlib for machine learning, GraphX for graph processing, and Spark Streaming[2].

Spark's ability to perform well on iterative algorithms makes it ideal for implementing Machine Learning Techniques as, at their vast majority, Machine Learning algorithms are based on iterative jobs. MLlib[3] is Apache Spark's scalable machine learning library and is developed as part of the Apache Spark Project. MLlib contains implementations of many algorithms and utilities for common Machine Learning techniques such as Clustering, Classification and Regression.

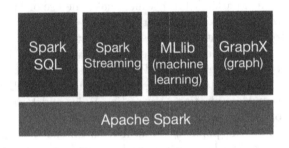

Fig. 1. Apache spark API stack

Apache Cassandra[4] is a free and open-source distributed database management system designed to handle large amounts of data across many commodity servers. It also has the ability to provide users with high availability with no single point of failure. Cassandra offers robust support for clusters spanning multiple data centers, with asynchronous master less replication allowing in this way, low latency operations for all clients. Its data structure model is implemented on a chord based Distributed Hash Table (DHT) and offers a hybrid database management system between key-value pairs as it is also column oriented. Cassandra also introduces CQL (Cassandra Query Language), a SQL-like alternative for addressing queries, which hides the implementation details of the data structures behind, while preserving load balancing and fault tolerance by replicating data in an automatic way.

---

[2] http://spark.apache.org/streaming/.

[3] http://spark.apache.org/mllib/.

[4] http://cassandra.apache.org/.

3   Machine Learning Algorithms

Based on Spark's built-in machine learning library called MLlib, we implement and experiment with six machine learning algorithms [16] on a given Point-Of-Sales (POS) data set comparing the results. Similar work considering POS datasets can be found in [1,2,5,9,11,15].

Initially, we start the experiments with Naive Bayes[5], a simple multiclass classification algorithm that can be trained very efficiently. It computes the conditional probability distribution of each feature given label, and then it applies Bayes' theorem so as to compute the conditional probability distribution of label given an observation and finally it uses it for prediction. In the same category, we experiment with linear SVM (Support Vector Machine)[6], which is a standard method for large-scale classification tasks that provides as output an SVM model. This model can be used for predicting whether a given data point can have a positive or negative outcome.

In clustering category, K-Means algorithm[7] is introduced, which is one of the most commonly used clustering algorithms that clusters the data points into a predefined number K of clusters. Then, it is combined with Gaussian Mixture Model, which represents a composite distribution whereby points are drawn from one of K Gaussian sub-distributions, each with its own probability model. Moreover, Expectation Maximization[8] is used as the learning clustering function which yields comprehensive results of not only the inferred topics but also the full training corpus and topic distributions for each document in it.

In frequent pattern mapping category, Association Rules[9] are utilized, which implement a parallel rule generation algorithm for constructing rules that have a single item as the consequent and FP-Growth [4] that consists of the following three steps. The first step of FP-Growth is to calculate item frequencies and identify frequent items. In following, the second step uses a suffix tree (FP-Tree) structure in order to encode transactions without generating candidate sets explicitly, which are usually expensive to be generated. As last step, the frequent item sets can be extracted from the FP-tree. In Spark's MLlib, a parallel version of FP-Growth called PFP [10] is implemented, which distributes the work of growing FP-Trees based on the suffices of transactions.

4   Describing the Dataset

Data analysis in business industry aims to extract knowledge from stored data to support business decisions. The dataset used for this experimental case study refers to consumer demand data from supermarket stores (POS). The data stored in Cassandra has the following form (Table 1):

---

[5] https://en.wikipedia.org/wiki/Naive_Bayes_classifier.
[6] https://en.wikipedia.org/wiki/Support_vector_machine.
[7] https://en.wikipedia.org/wiki/K-means_clustering.
[8] https://en.wikipedia.org/wiki/Expectation%E2%80%93maximization_algorithm.
[9] https://en.wikipedia.org/wiki/Association_rule_learning.

Table 1. Dataset description

Auto_Inc	Basket_ID	Date	Barcode	Sum_Units	Sum_Value
1	959980460	40938	520121904010	2	12,15
2	959980460	40938	212735100000	1	5
3	959980461	40938	356007034249	1	1,34
4	959980461	40938	520100404062	1	9,50

The Point-Of-Sales data relates to the transactions that have been made in the cash register of a supermarket. Each *Basket_ID* refers to a receipt. The *Date* field represents the date that the transaction (as an integer) has been taken place. The *Barcode* field refers to a unique product number, while *Sum_Units* are the number of units of the given barcode purchased. Also, the *Sum_Value* is the total value of those purchases. It is clear that the *Auto_Inc* field represents the record number in Cassandra and the primary key of the table.

5   Applying Machine Learning Algorithms to Data

As mentioned in the previous Sect. 3, for the experimental case study, several Machine Learning Algorithms from Spark's built-in library MLlib are used. In this Section, we will explain the appliance of these algorithms on the dataset as well as the possible uses of the extracted results.

5.1   Frequent-Pattern Mining

As being stated in MLlib main page, mining frequent items, itemsets, subsequences, or other sub-structures is usually among the first steps in the analysis of a large-scale dataset, which has been an active research topic in Data Mining for years. We have used Spark's parallel implementation of FP-Growth on our dataset in several ways so that we can extract knowledge and come to different type of conclusions. There are two basic (hyper-)parameters in FP-Growth implementation, namely:

1. *minSupport*: the minimum support for an itemset to be identified as frequent. For example, if an item appears 3 out of 5 transactions, then it has a support of $\frac{3}{5} = 0.6$.
2. *numPartitions*: the number of partitions used to distribute the work.

We experimented with these values as there is not an optimal value; the corresponding value depends on the nature of the problem. Initially, we run the algorithm with *Barcode* and *Sum_Units* columns of the table looking for the association between these two attributes. An example of the result form is shown below (Table 2) where all the items units as well as purchase frequencies (*Item* with *Barcode* 520131000607 purchased in 3 units twice etc.) are considered.

Table 2. Barcode and sum units columns

Barcode	Unit	Frequency
520131000607	3	2
520553310006	4	1
520126100000	8	1
2091199	1	547

These can be used so as to spot all the items that have frequently been purchased one, in pair or more.

Another way of using FP-Growth with a Point-Of-Sales dataset is by trying to associate the *Barcode* with the *Date* that the purchase is taken place. In this way, a pattern between certain items and the dates that they are sold the most is considered. The results of this experiment is shown below, representing the *Barcode* and the *Date* along with the frequency of this pair. As we can see from the following Table 3, the *Item* with *Barcode* 520131000607 has not been purchased on certain dates.

Table 3. Barcode with the date

Barcode	Unit	Frequency
520131000607	41337	7
520131000607	41338	7
520131000607	41339	0
520131000607	41340	10
520131000607	41341	9
520131000607	41342	5
520131000607	41343	0

5.2   K-Means Clustering

One of the most popular unsupervised machine learning and clustering algorithms is K-Means. Over the past decade, data have grown larger and larger and thus, the need of analysis over these corresponding data has risen. K-Means popularity has kept up with this growth because of its simplicity to be implemented and in following to be executed. K-Means is also used for completing a draft, pre-clustering procedure before a smarter, more sophisticated algorithm takes action.

All over the globe, we are generating millions of Gigabytes of data. A vast portion of these data, is Point-Of-Sale data. Point-Of-Sale (POS) and Point-Of-Purchase (POP) data deal with the time and place where a retail transaction is completed. Every individual on the retail/whole sale industry is generating daily this kind of data no matter the nature of business. Furthermore and without any doubt, K-Means is practical and thus, this kind of data can be processed by this algorithm.

Taken the above into consideration, K-Means could be combined with these data in order to produce useful conclusions. For the corresponding experiments, we have used with minimum variations, the data set we described above and we kept the *Product_ID* and the *Date* it was sold. Our purpose was to find patterns of the items that sold in specific dates. Due to the format of the values stored, we had to normalize the data.

Regarding the *Barcode* column, it is obvious that each code represents one single item. So, we have conducted a modulo operation dividing by the number of items belonging in the list, thus 9.412. The above procedure took part just for normalization purposes in order for the results to be translated in a more understandable form. We plotted using a scatter plot, where the X-axis represents days (from 0 to 59), and the Y-axis represents the IDs of the products. Their X-Y point projection represents a specific transaction.

As mentioned before, K-Means is an easy to be executed algorithm, thus we have utilized Spark's K-Means implementation from MLlib. It requires an input space separated file as well as an integer, representing the centers that should be found.

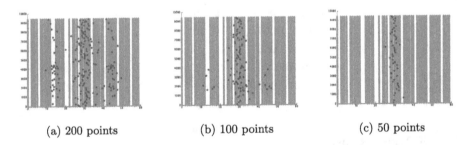

(a) 200 points   (b) 100 points   (c) 50 points

Fig. 2. K-Means results

Running the algorithm on the scattered plotted POS data, we have came across with the following finding; as we entered fewer center points, these were converged to the center of the data set as shown in Fig. 2. These K-Means points should be the starting point of other, more intelligent, clustering algorithms to be focused on. We are convinced that these data are totally exploitable by businesses and organizations and we are into a more in depth analysis.

5.3   Naive Bayes

Naive Bayes is maybe the simplest classification algorithm. Belonging in multi-class category, the classification of any object in one of two or more classes is made possible. Regarding Spark, Naive Bayes can be trained with efficiency by just one pass on the training data provided. After this procedure, the model can receive any object given and in following predict in which class it belongs.

In the following Table 4, there is a sample training data set with *POS* data for the Naive Bayes algorithm. Classes were created by running the *K*-Means clustering algorithm in the *Basket_ID* of our original POS data set.

Table 4. Sample training data set

Class number	Barcode_1	Barcode_2	Barcode_3
0.0	520423911961	520102809110	520423930862
0.0	210452200000	520423911961	210757100000
0.0	520102809110	210452200000	520423930862
1.0	520131469697	520423911961	210757100000
1.0	210452200000	520102809110	520423930862
2.0	520423911961	210757100000	210452200000
2.0	520423930862	520102809110	520423911961

After training Naive Bayes algorithm with the previous data set, we have predicted the class of some other baskets found on our data set. Next Table 5 shows the barcodes of the basket's items as well as the class that Naive Bayes predicted the basket will belong to.

Table 5. Class prediction using naive bayes

Barcode_1	Barcode_2	Barcode_3	Class_Prediction
210452200000	520102809110	210757100000	1.0
210757100000	520423911961	520102809110	1.0

In the real sales world, Naive Bayes can be used in order to classify customers. Depending on what customers bought and using the predictive model this algorithm provides, companies can make special offers on a group of buyers. Furthermore, companies can even provide some real-time discount for an additional product that a specific customer could buy, depending on what other customers with same profile have bought.

5.4   Linear Support Vector Machines (SVM)

The biggest difference between SVM and Naive Bayes is the number of classes in which algorithms can classify the objects. As SVM belong in the Binary Classification algorithm, thus it can classify objects into only two categories. Despite that, SVM are more accurate as Spark gives the option of running through training data more than one time. Another one difference is the mathematical model of prediction behind SVM.

SVM algorithm was trained with the POS data set as shown above in Table 4. The data were classified again using K-Means algorithm but in this Table, there are only two categories because SVM is a Binary Classification Algorithm as mentioned above. After training the SVM algorithm with the training data, we randomly chose some baskets from our POS data set in order to classify them. In Table 6, the baskets as well as the classes that SVM algorithm predicted they belong to, are presented.

Table 6. Class prediction using SVMs algorithm

Barcode_1	Barcode_2	Barcode_3	Class_Prediction
210452200000	520102809110	520423911961	0.0
210452200000	520423911961	210757100000	0.0

In general, SVM may help companies in order for them to take some boolean decisions as they classify data in just two categories. To conclude, Classifying Algorithms can provide companies with useful data, most of them are easy to use and the companies have just to choose between accuracy and time as the more accurate an algorithm could be, the slower the training will be as it will need more iterations on the training data.

6   Conclusion and Future Steps

In this paper we presented a survey of machine learning algorithms implemented on Apache Spark over DHT-structures. We experimented across a POS dataset that had been stored in Cassandra. The experiments conducted with some of the most influential algorithms that have been widely used in the machine learning community; these are Naive Bayes, SVM, K-Means, Expectation Maximization, Association Rules and Parallel FP-Growth.

As future work, we plan to introduce another similar significant problem, that is data mining on cloud computing (e.g. Spark) over Hierarchical-based Tree Structures. This survey of machine learning algorithms on Spark over Hierarchical-based Tree Structures will be based on manuscripts like [6, 7, 12–14].

References

1. Cho, Y.H., Kim, J.K., Kim, S.H.: A personalized recommender system based on web usage mining and decision tree induction. Expert Syst. Appl. **23**(3), 329–342 (2002)
2. Dickson, P.R., Sawyer, A.G.: The price knowledge and search of supermarket shoppers. J. Mark. **54**, 42–53 (1990)
3. Gourgaris, P., Kanavos, A., Makris, C., Perrakis, G.: Review-based entity-ranking refinement. In: Proceedings of the 11th International Conference on Web Information Systems and Technologies (WEBIST), pp. 402–410 (2015)
4. Han, J., Pei, J., Yin, Y.: Mining frequent patterns without candidate generation. In: Proceedings of the 2000 ACM SIGMOD International Conference on Management of Data, pp. 1–12 (2000)
5. Iakovou, S.A., Kanavos, A., Tsakalidis, A.: Customer behaviour analysis for recommendation of supermarket ware. In: Iliadis, L., Maglogiannis, I. (eds.) AIAI 2016. IAICT, vol. 475, pp. 471–480. Springer, Cham (2016). doi:10.1007/978-3-319-44944-9_41
6. Jagadish, H.V., Ooi, B.C., Tan, K., Vu, Q.H., Zhang, R.: Speeding up search in peer-to-peer networks with a multi-way tree structure. In: Proceedings of the ACM SIGMOD International Conference on Management of Data, pp. 1–12 (2006)
7. Jagadish, H.V., Ooi, B.C., Vu, Q.H.: BATON: A balanced tree structure for peer-to-peer networks. In: Proceedings of the 31st International Conference on Very Large Data Bases (VLDB), pp. 661–672 (2005)
8. Kanavos, A., Kafeza, E., Makris, C.: Can we rank emotions? A brand love ranking system for emotional terms. In: 2015 IEEE International Congress on Big Data, pp. 71–78 (2015)
9. Leskovec, J., Adamic, L.A., Huberman, B.A.: The dynamics of viral marketing. ACM Trans. Web (TWEB) **1**(1), 5 (2007)
10. Li, H., Wang, Y., Zhang, D., Zhang, M., Chang, E.Y.: PFP: parallel FP-growth for query recommendation. In: Proceedings of the 2008 ACM Conference on Recommender Systems (RecSys), pp. 107–114 (2008)
11. Pennacchioli, D., Coscia, M., Rinzivillo, S., Pedreschi, D., Giannotti, F.: Explaining the product range effect in purchase data. In: Proceedings of the 2013 IEEE International Conference on Big Data, pp. 648–656 (2013)
12. Sioutas, S., Papaloukopoulos, G., Sakkopoulos, E., Tsichlas, K., Manolopoulos, Y.: A novel distributed P2P simulator architecture: D-P2P-sim. In: Proceedings of the 18th ACM Conference on Information and Knowledge Management (CIKM), pp. 2069–2070 (2009)
13. Sioutas, S., Papaloukopoulos, G., Sakkopoulos, E., Tsichlas, K., Manolopoulos, Y., Triantafillou, P.: Brief announcement: Art: Sub-logarithmic decentralized range query processing with probabilistic guarantees. In: Proceedings of the 29th Annual ACM Symposium on Principles of Distributed Computing (PODC), pp. 118–119 (2010)
14. Sioutas, S., Triantafillou, P., Papaloukopoulos, G., Sakkopoulos, E., Tsichlas, K., Manolopoulos, Y.: ART: Sub-logarithmic decentralized range query processing with probabilistic guarantees. Distrib. Parallel Databases **31**(1), 71–109 (2013)
15. Weng, S., Liu, M.: Feature-based recommendations for one-to-one marketing. Expert Syst. Appl. **26**(4), 493–508 (2004)
16. Witten, I.H., Frank, E., Hall, M.A., Pal, C.J.: Data Mining: Practical Machine Learning Tools and Techniques. Morgan Kaufmann, San Francisco (2016)

17. Wu, X., Kumar, V., Quinlan, J.R., Ghosh, J., Yang, Q., Motoda, H., McLachlan, G.J., Ng, A.F.M., Liu, B., Yu, P.S., Zhou, Z., Steinbach, M., Hand, D.J., Steinberg, D.: Top 10 algorithms in data mining. Knowl. In. Syst. **14**(1), 1–37 (2008)
18. Zaharia, M., Chowdhury, M., Das, T., Dave, A., Ma, J., McCauly, M., Franklin, M.J., Shenker, S., Stoica, I.: Resilient distributed datasets: A fault-tolerant abstraction for in-memory cluster computing. In: Proceedings of the 9th USENIX Symposium on Networked Systems Design and Implementation (NSDI), pp. 15–28 (2012)
19. Zaharia, M., Chowdhury, M., Franklin, M.J., Shenker, S., Stoica, I.: Spark: Cluster computing with working sets. In: 2nd USENIX Workshop on Hot Topics in Cloud Computing (HotCloud) (2010)

Scaling Out to Become Doctrinal

Yannis Panagis[1]([⊠]) and Evangelos Sakkopoulos[2]

[1] iCourts Centre of Excellence for International Courts, University of Copenhagen,
1455 Copenhagen, Denmark
ioannis.panagis@jur.ku.dk
[2] Hellenic Open University, Patra, Greece
evangelos.sakkopoulos@ac.eap.gr

Abstract. International courts are often prolific and produce a huge amount of decisions per year which makes it extremely difficult both for researchers and practitioners to follow. It would be thus convenient for the legal researchers to be given the ability to get an idea of the topics that are dealt with in the judgments produced by the courts, without having to read through the judgments. This is exactly a use case for topic modeling, however, the volume of data is such that calls for an out-of-core solution. In this paper we are experimenting in this direction by using the data from two major, large international courts. We thus, experiment with topic modeling in Big Data architectures backed by a MapReduce framework. We demonstrate both the feasibility of our approach and the accuracy of the produced topic models that manage to outline very well the development of the subject matters of the courts under study.

Keywords: Big Data · MapReduce · European Court of Justice · European Court of Human Rights · Latent Dirichlet Allocation

1 Introduction

Researchers in law are often faced with the task of studying the case-law of the court(s) in question, for the purposes of their research. It is then that they are confronted with the problem of going through vast case-law collections and several years of discourse development. Therefore, it is an accepted practice for the researchers to only explore a small subset of cases (see the discussion in [10]), or to keep track of the case-law by reading legal summaries.

Machine Learning has at the same time been providing for several years now with methods to summarize text and to extract topics from text. The latter area is widely known as *topic modeling*. With the availability of topic modeling techniques on one hand and the case-law text collections on the other, it is intriguing to see the applicability of topic modeling in case-law and this is the driving force for this paper. Albeit, the topic modeling methodologies are extremely resource intensive and this attribute renders them difficult to manage in a monolithic computing environment, given the size of the case-law of major courts.

© Springer International Publishing AG 2017
T. Sellis and K. Oikonomou (Eds.): ALGOCLOUD 2016, LNCS 10230, pp. 157–168, 2017.
DOI: 10.1007/978-3-319-57045-7_10

Therefore, we have to seek for other solutions that scale out well and can handle large text collections. The MapReduce framework [9] is ideal for that purpose because it can handle very large amounts of data and can solve the computation bottleneck, by dispersing the computations among a group of computers in a network.

In this paper we use MapReduce and apply topic modeling, and more specifically *Latent Dirichlet Allocation*(LDA) [2] to the case-law of two major international courts, the *European Court of Justice* and the *European Court of Human Rights*. Our goal is twofold: (1) test the applicability of topic modeling to model case-law, especially how feasible it is to detect the topics that the courts were dealing with through their lifespan (2) check how well does topic modeling work in practice, in a MapReduce setting and if it is practical to apply it to this category of problems or, in other words, if it is feasible to scale out our infrastructure in order to learn the doctrine of international courts.

The rest of the paper is organized as follows: Sect. 2 gives a background to the courts and related work as well as a presentation of topic modeling and MapReduce. Section 3 describes our dataset and presents the experiment we conducted. In Sect. 4 we evaluate our approach qualitatively and quantitatively and finally, Sect. 5 concludes the discussion.

2   Background

2.1   Presentation of the Courts and Related Work in the Legal Field

As mentioned in the introduction, in this paper we are using the case-law, i.e. the set of judgments, of the European Court of Justice, hence ECJ, and of the European Court of Human Rights, hence ECHR. Both courts are similar and with similar history. They were founded in the 1950s, they have international jurisdiction across Europe, they underwent major changes at the end of 90s-beginning of 00s, and today they produce a large number of judgments every year, having today around 12.000 judgments (ECJ) and 18.000 (ECHR).

Due to the amount of the produced case-law, the courts have frequently been the object of qualitative research in, among others, Legal and Political Science. A path that is often used to study the courts qualitatively is that of *network analysis*. Network analysis in the case of case-law and law in general, entails building a directed citation graph of citations to either precedent or articles, instruments, paragraphs and son on. After the network has been built one seeks to detect important cases with the use of network importance metrics such as HITS [12] and PageRank [5]. Examples of this approach are [11,23,25]. Others apply community detection to find interesting case clusters [16]. In many occasions researchers use network metrics as a method to complement their analysis of case-law, see for example [14,20,21].

In the machine learning domain and more specifically machine learning for texts, there are, to the best of our knowledge, only very few papers that deal with case-law. Mendoza and Fürnkranz [13] present a model to learn multilabel classification for the documents of the European Union (including case law of the ECJ). Topic modeling has been used by Winkels et al. [24] and Boer and Winkels [4], to build a recommender system for case-law. Panagis and Šadl [19], use LDA complementary to network analysis, in order to create topic models from the top cited paragraphs of the cases of the ECJ about *european citizenship*. The topics are used to see what areas are discussed in the relevant case-law on that subject and whether the subject of the citation is the area itself or some legal principle. In a slightly different line of research, Panagis et al. [18] deploy Non-Negative Matrix Factorization to uncover the topics in the various periods of the two courts and to study the topic development.

The paper that is most similar to ours is that of Nagwani [17]. In this paper the author uses LDA as a step in a machine learning workflow which aims at summarizing case-law. Nagwani also uses MapReduce to improve performance, his dataset comprises, however, of only 4000 judgments[1].

2.2   Topic Modeling

Topic modeling is the branch of machine learning that aims at constructing a model that infers the topics being discussed in the underlying text collection. The field was revolutionized by the development of *Latent Dirichlet Allocation,* by Blei et al. [2], where LDA subsequently gave rise to a whole family of methods. In the following paragraphs we will give some basic notions of the LDA. The reader is referred to [3] for an excellent exposition of the field.

LDA is special in the way that it model the topics in text in the sense that it treats texts as unordered collections of words (bag of words) and tries to go through these collection iteratively to infer the topic distributions. Topic inference can then be described as an iteration of the following two steps [3].

We will now give some notation to explain in more detail the idea and the methods behind LDA. Let D denote the corpus consisting of N documents and let K be the number of topics. The topics are $\beta_{1:K}$, where each β_k is a distribution over the vocabulary. The topic proportions for the dth document are θ_d, where $\theta_{d,k}$ is the topic proportion for topic k in document d. The topic assignments for the dth document are z_d and $z_{d,n}$ is the topic assignment for the nth word in d. Finally, the set of words for document d is denoted as w_d, where $w_{d,n}$ is the nth word in document d, which is an element from the fixed vocabulary.

The procedure behind the LDA is shown in Fig. 1. It can further be described as a sequence of the following steps [2]

1. Randomly choose a distribution over topics hence inferring β. This distribution is Dirichlet with prior η.
2. Choose θ from a Dirichlet distribution with prior α.
3. For each word in the document:
   (a) Randomly choose a topic from the distribution over topics in step 2, above,
   (b) Randomly choose a word from the corresponding distribution over the vocabulary expressed as $p(w_n|z_n\beta)$.

---

[1] https://archive.ics.uci.edu/ml/datasets/Legal+Case+Reports.

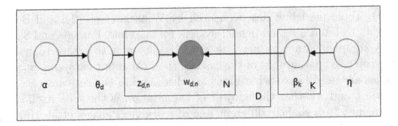

Fig. 1. Graphical model representation of LDA.

This procedure is modeled by a set of equations, the presentation of which is beyond the scope of this paper. We note though that in order to estimate the parameters in the all the distributions used for the sampling procedures, one should maximize the log-likelihood of the probability to observe the corpus with certain word-topic and document-topic distributions. This log-likelihood is expressed as:

$$l(D|\alpha, \beta) = \sum_{d=1}^{M} \log p(w_d|\alpha, \beta). \qquad (1)$$

Computing a solution to (1) is not a tractable problem. Therefore, a set of techniques have been developed to compute an approximation to it. The models in this paper were built with the use of the *Variational Expectation Maximization* technique, which is also implemented in Spark[2] [7].

2.3   MapReduce

MapReduce [8] is a popular programming model that is also appropriate for working with big data. The data processing in MapReduce is based on input data partitioning; the partitioned data is processed by a number of tasks executed in many distributed nodes. There exist two major task categories called *Map* and *Reduce* respectively. Given input data, a *Map* function processes the data and outputs key-value pairs. Subsequently, the data are shuffled by a *Shuffle* process, key-value pairs are grouped and then each group is sent to the corresponding Reduce task. A user can define his own Map and Reduce functions depending on the purpose of his application. The input and output of these functions are simplified as key-value pairs. This interface provides an abstraction level to the programmers, where they can concentrate on solving the actual computing problem and all the details of distribution, scheduling and execution are handled by the model implementation. The architecture of MapReduce model is depicted in Fig. 2. The figure shows a simplified instance of a MapReduce cluster. The cluster consists of four computing units, or *nodes*, one of which assumes the

---

[2] Our reference to Spark is premature at this point. For a more complete discussion see Sect. 2.3.

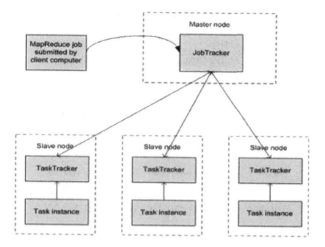

Fig. 2. Illustration of the MapReduce architecture.

role of the *master* and the rest are called *slaves*. The master undertakes task scheduling, communication and execution, while the slave nodes are executing the computing tasks.

One open source implementation of the MapReduce model is Hadoop [1]. Hadoop has further been improved by Spark [26] in order to provide more efficient data communication between the computers constituting the cluster, without the need to replicate the data in the Hadoop file system. This allows execution of computing tasks like machine learning with a speedup that can reach up to 10 times. Moreover, Spark includes a machine learning library, called MLLib, with bindings in Scala, Java, Python and R. This library includes, among others, an implementation for LDA that makes use of Variational Expectation Maximization, since, according to the authors [7], it provides for a natural implementation using GraphX, the graph processing library of Spark.

3   Data and Experiment

Our dataset consists of judgment texts from the ECJ and the ECHR that cover the period from the start of both courts until the end of 2014. Even though the courts publish in many languages, the language of the texts is English in both cases to facilitate our analysis. The texts were obtained from HUDOC[3] in the case of ECHR and from EUR-Lex[4] in the case of the ECJ.

3.1   Data Preparation

Figure 3 shows the processing workflow we applied in the paper. First, the texts were obtained from the the in-house available, iCourts' database of international

---

[3] http://hudoc.echr.coe.int/eng.
[4] http://eur-lex.europa.eu.

Table 1. The number of cases and number of tokens per court and period.

Periods	CJEU		ECHR	
	Cases	Tokens	Cases	Tokens
1950–60	358	951.643	12	178.553
1970	901	2.305.177	31	465.920
1980	1.771	5.443.995	192	1.594.448
1990	2.143	8.351.596	929	7.227.337
2000	1.358	7.549.826	1.921	11.349.603
2005	1.415	8.607.960	4.690	25.764.102
2010-14	989	6.240.242	2.791	22.715.120
Totals	8.935	39.450.439	10.566	69.295.083

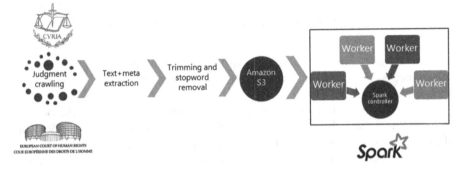

Fig. 3. The processing workflow.

courts[5]. The database includes not only the texts themselves but also metadata about the judgments, such as date, article, subject matter and so on. We, then, divided the texts into periods. The periods we used are 50–60, 70, 80, 90, 2000–05, 2005–10 and 2010–14. The motivation behind this division is to divide the collection per decade with the following exceptions: The 50's and the 60's were merged because both courts had very few judgments in the 50's–60's. Furthermore, during the 00's, both courts experienced a blast in number of cases decided per year, and this had as a result that the Spark computations took extremely long time to finish which made us decide to split it into two five-year periods. In the case of ECJ some very few judgments are not available in HTML, which we did not include for the sake of simplicity. Hence, the total number of texts from both courts is 19.501.

Since the judgments of the ECHR are mainly divided into, facts, law, decision and dissenting opinion in the next step we applied a trimming heuristic to the texts of ECHR after 2000 in order to further reduce the size of the collection and

---

[5] http://www.icourts.dk.

to concentrate to the main part of the case. Therefore, we kept the last part of the judgment which corresponds to the law section of it and we removed, where possible, the dissenting opinions. The number of files per period and the corresponding number of tokens per period after trimming was applied, are shown in Table 1. After trimming was applied, we ran stopword removal, using a standard stopword list for English[6], adding a few court specific words to the existing list, for example words like "court", "ecj", "echr", "article", etc.

The next step entailed transferring files to an Amazon S3 drive and from there they were made available to a Spark cluster for further processing. The cluster we used was provided by Databricks[7] and consisted of a controller and 8 Amazon EC2 r3.2xlarge instances.

3.2   Experiment

Experimental Details. For each court and period we ran the LDA implementation that is available by MLLib and by using Scala notebooks. As described in Sect. 2.2 we used the Variational Expectation Minimization technique. The advantage of this technique is that it converges fast and produces relatively meaningful results.

Table 2. The number of topics per period.

Periods	CJEU	ECHR
1950–60	10	15
1970	10	15
1980	20	20
1990	20	25
2000	30	30
2005	30	30
2010–14	25	25

For each court-period instance, we selected a different number of topics, k. The selection was made empirically knowing a priori that the courts had a small number of cases and a rather narrow area in their earlier period, 1950–70, and then they opened up both in terms of the number of cases they decided and in terms of topics. This is especially true during the 2000s, when the volume of judgments and the different topics increased dramatically, but we can observe a slight drop during the latest years. This observation is validated more rigorously in [18] and is justified by manual examination of our topics with different values of k. The different number of topics per court period is shown in Table 2.

---

[6] Available here: http://ir.dcs.gla.ac.uk/resources/linguistic_utils/stop_words.
[7] http://www.databricks.com.

Table 3. Representative topic words per court and period.

Periods	CJEU	ECHR
1950–60	Regulations, **Staff**, Equalization, **Undertakings**, **Scrap**, Costs, **Regulation**, Prices	Cases, Procedure, Legal, Fact, Detention, Criminal, **Ringeisen**, Public, Freedom, Time, Trial, Reasons, Arrest, **Belgian**, **Education**, Convention, Rights
1970	Costs, National, Provisions, Legislation, **Benefits**, Products, **Tariff**, **Customs**, Compensatory	Convention, Freedom, Rights, Detention, Time, Family, Child, Rights, Government, Compensation, Education, **Language**, König
1980	Question, Costs, National, **Pension**, Benefits, Production, Value, **Directive**, **Milk**	Criminal, Procedure, Compensation, Civil, Breach, Government, Time, Hearing, Authorities, Administrative, Courts, **Child**, **Family**, **United**, **Kingdom**
1990	Hearing, Provisions, Costs, Trade, Products, Customs, Pension, Benefits, Social, Legislation	Hearing, Convention, **Turkey**, Applicants, Public, Documents, Criminal, Cassation, Offence, **Military**, **Service**, Greek, Religious, Police, Officers, Force, **Property**, Land, Child, **Detention**, Civil
2000	Instance, Measures, **Contract**, Government, Products, Services, **Goods**, **Workers** Time, Customs, **Principle**, **Legal** Provision, Measures, **Effect**, Established, Right, Provide, Appeal, Undertakings	Convention, Criminal, Procedure, Police, Prison, Right, Domestic, Authorities, Russian, **Detention**, Evidence, **Pecuniary**, **Damage**, Government, **Time**, Authorities, **Turkey**, **Security**, Violation, Convention, **Property**, Land, Applicants
2005	Rights, Property, Period, **Obligations**, Customs, Employment, **Workers**, Provisions, **Freedom**, **Establishment**, Goods, **Trade**, **Mark**, Company, Capital, Procedure, Services, Person	**Length**, Police **Investigation**, Convention, Complaint, Constitutional, **Poland**, Turkey, Pecuniary, Damage, Violation, Detention, Expenses, Applicants, Protocol, Enforcement, Domestic, Violation, Rights, **Family**
2010–14	Company, Treatment, Product, **Social**, **Security**, Leave, Appeal, Notice, Rules, Persons, Workers, Costs, Customs, Services, Trade, Mark	Investigation, Police, Criminal, Civil, Compensation, **Asylum**, Convention, Russia, **Just satisfaction**, Applicants, Rate, **Settlement**, Length, **Medical treatment**, Evidence, Expenses, **Prison**, **Conditions**, Detention, Pecuniary, Damage

Results. The output of topic modeling is a set of topics and each topic is described by a set of words, where each word is ranked with the probability to describe the specific topic. Thus, for each court and period we collected the top-10 words per topic. Due to the inherent difficulty to present all topics here, we chose to pick some representative words from the topic words of each period and present them in Table 3. The words in bold denote words that can be considered indicative of the judgments of the specific period.

4   Evaluation

The evaluation of the produced models is twofold. First, we evaluate them from a content perspective and then, the performance of the LDA cluster is evaluated versus an established centralized implementation.

From a content perspective, the results give a broad picture of the subject dealt from both courts across time. The picture we get is very coarse but this is quite normal taking into account the breadth of the subjects a court has dealt with, along time. Beginning from the ECJ, we can see that during the first years the model picks up terms like *Undertakings* and *Regulation*. This is absolutely normal since the ECJ was built as the Court of the Coal and Steel Community and it was dealing initially with subjects related to the european market of steel. Later on as it opens up to become a pan-european court we see terms like *Benefits* and *Customs* (70's) and even later in the 80's *Directive* and *Milk*. The latter marks a fundamental case in EU-law, namely the case *120/86 - Mulder v Minister van Landbouw en Visserig*, a case that helped to develop the doctrine of *legitimate expectations* in EU Law [22]. In the 2000–05 period, we can see concepts like *Direct Effect* and *Worker* both of which were established in previous periods but which started becoming more popular in that period, see [6,19]. In the latter periods the ECJ addresses concepts like *Freedom of Establishment*, *Social Security* and *Trade Marks*, all of which have been the subject of the Court's case-law the previous years.

The results from the ECHR are equally, interesting. The topics start with a reference to *Belgian* and *Language* as well as case names like *Ringeisen*. Both are well known cases from the early case-law of the Court. The first is case 1474/62 also known as *Belgian Linguistics*, a case about the right to education in Belgium. In the 1980's the subjects of the Court shift a bit towards the rights of the child and we also observe an entry United Kingdom due to a set of cases against the United Kingdom filed in this period. In the 1990s we see still the rights of the child to stay popular but we also observe terms like *Turkey*, *Security*, *Greece* and *Property* to crop up. This is explained by a batch of cases against Turkey involving the Turkish National Security Court and a batch of cases against Greece regarding land and properties, see e.g. *Iatridis v. Greece* or other human rights violations during military service, e.g. *Grigoriades v. Greece*.

As the Court moves in the 2000's we see other term surfacing like *Russian* and *Pecuniary Damage*. In 1999 the Court started examining cases from the east european countries and a lot of them at least in the initial period were against Russia. Furthermore, the Court started examining several applications regarding pre-trial detention and pecuniary damage. The interesting results continue in the 2005 period where we note the appearance of *Poland*, a country against which a lot of cases regarding the length of proceedings were brought, as well as terms like *Family* (together with rights) again with several cases in the subject. Pecuniary damage remains as a subject in this period, too. Finally, we can observe that in the last period we are having cases regarding *asylum seekers*, and *prison conditions* often associated with *medical treatment*, which is also what one can find in the relevant case-law.

A final observation for both courts is that the induced topics contain a lot of procedural terms like *procedure, provisions, costs, compensation* and so on, which is quite normal since the decision documents are always about some procedural topic.

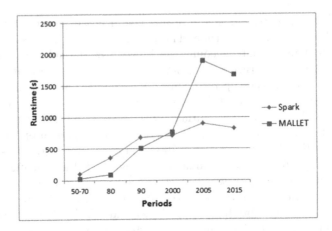

Fig. 4. Execution time comparison between the MLLib implementation of LDA and MALLET.

For the second part of the evaluation, we will compare the execution times of the Spark LDA implementation with the execution times needed for building a topic model with MALLET [15], a widely used topic modeling library, that runs in a single computer. We constructed with MALLET topic models for the ECHR, with the same values of k as with Spark, only now the model makes a 1000 iterations before it exits. We ran MALLET on a Windows 2008R2 Server with 4 CPUs and 16 GB RAM. The execution times comparison is shown in Fig. 4.

One has to note before drawing conclusions that the two implementations are not directly comparable since they are running different estimation techniques (Collapsed Gibbs sampling vs. Variational Expectation Maximization) but the configurations are close to what we would use in practice: It is recommended to use 1000 iterations to get meaningful results in MALLET and 120 to 200 iteration in the MLLib implementation, result in a reasonable execution time. The topics that we get from MALLET, though, are more meaningful on purely qualitative criteria. We see from Fig. 4 that the MLLib implementation runs almost twice as fast as MALLET in the most recent collections where the collection sizes grow significantly. For small collections MALLET runs faster, whereas MLLib is hitting a bottleneck in the communication overhead between nodes in the cluster. Overall, one can conclude that it is still reasonable to run MALLET with collections that have the size of our collection, since running the topic modeling in Spark for 1000 iterations as well, would result in larger execution times by an order of magnitude. Nevertheless, it would be interesting to compare MALLET with MLLib on e.g. the whole collection of ours, this, however, would require a totally different research setup, where one would not seek to study the development, hence topic models, of the court(s) by period but rather study the concepts developed in the entire court(s).

5   Conclusions

In this paper we have experimented with a very challenging task that of using a big data machine learning framework as a tool to study the case-law development of two major international courts. The results are encouraging and are showing that the topic model can pick some important topics that were examined by both courts at the same period. We have also seen that a centralized machine learning model is still a viable option for the given collection sizes. It is interesting in terms of applicability of the methods discussed here, to compare the MapReduce implementation with the single computer solution, when the entire corpus is used to build the topic models. This would require to specify a very large number of topics though, so that the model captures as many more concepts as possible.

What the models do not seem to capture are the legal, doctrinal concepts that are developed in the texts, like for example *direct effect* or *principle of subsidiarity*. In the absence of a controlled vocabulary for international courts, a possible method to tackle this problem would be to use techniques for terminology extraction and use the extracted terms as tokens, instead of the individual words. We plan to work towards this research direction, as well as to compare in a more rigorous way, the models generated by LDA with those produced by Non-Negative Matrix Factorization.

Acknowledgements. Yannis Panagis is partially supported by the project "From Dogma to Data: Exploring How Case Law Evolves" Danish Council for Independent Research project ID: DFF-4003-00164.

References

1. Apache Foundation: The hadoop project. http://hadoop.apache.org
2. Blei, D.M., Ng, A.Y., Jordan, M.I.: Latent Dirichlet allocation. J. Mach. Learn. Res. **3**, 993–1022 (2003)
3. Blei, D.M.: Probabilistic topic models. Commun. ACM **55**(4), 77–84 (2012)
4. Boer, A., Winkels, R.: Making a cold start in legal recommendation: an experiment. In: JURIX. Frontiers in Artificial Intelligence and Applications, vol. 294, pp. 131–136. IOS Press (2016)
5. Brin, S., Page, L.: Reprint of: the anatomy of a large-scale hypertextual web search engine. Comput. Netw. **56**(18), 3825–3833 (2012)
6. Craig, P., De Búrca, G.: EU Law: Text, Cases, and Materials. Oxford University Press, Oxford (2011)
7. DataBricks: Topic modeling with LDA: MLLib meets graphx. https://databricks.com/blog/2015/03/25/topic-modeling-with-lda-mllib-meets-graphx.html
8. Dean, J., Ghemawat, S.: MapReduce: a flexible data processing tool. Commun. ACM **53**, 72–77 (2010). http://dx.doi.org/10.1145/1629175.1629198
9. Dean, J., Ghemawat, S.: MapReduce: simplified data processing on large clusters. In: 6th Symposium on Operating System Design and Implementation (OSDI 2004), San Francisco, California, USA, 6–8 December 2004, pp. 137–150 (2004)
10. van Gestel, R., Micklitz, H.W.: Why methods matter in European legal scholarship. Eur. Law J. **20**(3), 292–316 (2014)

11. Guillaume, G.: The use of precedents by international judges and arbitrators. J. Int. Dispute Settl. **2**(1), 5–23 (2011)
12. Kleinberg, J.M.: Authoritative sources in a hyperlinked environment. J. ACM **46**(5), 604–632 (1999)
13. Loza Mencía, E., Fürnkranz, J.: Efficient pairwise multilabel classification for large-scale problems in the legal domain. In: Daelemans, W., Goethals, B., Morik, K. (eds.) ECML PKDD 2008. LNCS (LNAI), vol. 5212, pp. 50–65. Springer, Heidelberg (2008). doi:10.1007/978-3-540-87481-2_4
14. Lupu, Y., Voeten, E.: Precedent in international courts: a network analysis of case citations by the European Court of Human Rights. Br. J. Polit. Sci. **42**(02), 413–439 (2012)
15. McCallum, A.K.: Mallet: a machine learning for language toolkit (2002). http://mallet.cs.umass.edu
16. Mirshahvalad, A., Lindholm, J., Derlen, M., Rosvall, M.: Significant communities in large sparse networks. PloS One **7**(3), e33721 (2012)
17. Nagwani, N.K.: Summarizing large text collection using topic modeling and clustering based on MapReduce framework. J. Big Data **2**(1), 6 (2015)
18. Panagis, Y., Christensen, M.L., Sadl, U.: On top of topics: leveraging topic modeling to study the dynamic case-law of international courts. In: JURIX. Frontiers in Artificial Intelligence and Applications, vol. 294, pp. 161–166. IOS Press (2016)
19. Panagis, Y., Sadl, U.: The force of EU case law: a multi-dimensional study of case citations. In: JURIX. Frontiers in Artificial Intelligence and Applications, vol. 279, pp. 71–80. IOS Press (2015)
20. Sadl, U., Madsen, M.R.: A "selfie" from Luxembourg: the court of justice and the fabrication of the pre-accession case-law dossiers. Columbia J. Eur. Law **22**(2), 327–354 (2016)
21. Sadl, U., Panagis, Y.: What is a leading case in EU law? An empirical analysis. Eur. Law Rev. **40**(1), 15–34 (2015)
22. Sharpston, E.: European community law and the doctrine of legitimate expectations: how legitimate, and for whom. Northwest. J. Int. Law Bus. **11**, 87 (1990)
23. Tarissan, F., Nollez-Goldbach, R.: Analysing the first case of the international criminal court from a network-science perspective. J. Complex Netw. **4**(4), 1–19 (2016). https://doi.org/10.1093/comnet/cnw002
24. Winkels, R., Boer, A., Vredebregt, B., van Someren, A.: Towards a legal recommender system. In: JURIX. Frontiers in Artificial Intelligence and Applications, vol. 271, pp. 169–178. IOS Press (2014)
25. Winkels, R., de Ruyter, J., Kroese, H.: Determining authority of Dutch case law. In: Atkinson, K. (ed.) Legal Knowledge and Information Systems (JURIX). Frontiers in Artificial Intelligence and Applications, vol. 235, pp. 103–112. IOS Press (2011)
26. Zaharia, M., Chowdhury, M., Franklin, M.J., Shenker, S., Stoica, I.: Spark: cluster computing with working sets. In: HotCloud. USENIX Association (2010)

A Survey on Big Data and Collective Intelligence

Ioannis Karydis[1]([⊠]), Spyros Sioutas[1], Markos Avlonitis[1], Phivos Mylonas[1],
and Andreas Kanavos[2]

[1] Ionian University, 49132 Kerkyra, Greece
{karydis,sioutas,avlon,fmylonas}@ionio.gr
[2] University of Patras, 26504 Patra, Greece
kanavos@ceid.upatras.gr

Abstract. The creation and accumulation of Big Data is a fact for a plethora of scenarios nowadays. Sources such as the ever-increasing diversity sensors as well as the content created by humans have contributed to the Big Data's enormous size and unique characteristics. Making sense of these data has primarily rested upon Big Data analysis algorithms. Still, in one too many cases the effectiveness of these algorithms is hampered by the very nature of Big Data: analogue, noisy, implicit, and ambiguous. Enter Collective Intelligence: the capability of interconnected intelligences achieving ameliorated results in activities than each of the single intelligences creating the collective solely would. Accordingly, this work presents existing research on Big Data and Collective Intelligence. The work is concluded with the presentation of the challenges and perspectives of the common ground between the directions of Big Data and Collective Intelligence.

Keywords: Big Data · Collective intelligence · Cloud computing · NoSQL · Crowdsourcing · Distributed systems · Synergetic networks

1 Introduction

The capability to create and store information nowadays is unparalleled. The gargantuan plethora of sources that leads to information of varying type, quality and consistency, large volume, creation rate per time unit has been identified as of 2001 [16]. Such data, also known affectively as Big Data, are currently at the order of tens of pebibytes [14] with increasing tendency, while their management and analysis are unsurprisingly a prominent research direction [13].

The continuously expansive ubiquitousness of cheap, mobile, network capable, multi-sensory, power-efficient processing capability is in part responsible for the aforementioned creation of Big Data (sets). In relation to the rate of evolution of the processing capability, storage capability of the produced information advanced somewhat less and thus the notion of streaming data with transient character. Another, less commonly identified factor for the emergence of Big Data, is the change of paradigm in content creation wherein willingness to contribute, domain expertise and access to content sharing methods became no

© Springer International Publishing AG 2017
T. Sellis and K. Oikonomou (Eds.): ALGOCLOUD 2016, LNCS 10230, pp. 169–181, 2017.
DOI: 10.1007/978-3-319-57045-7_11

longer a capability/privilege of few people but a widespread common ground, especially under the auspices of "web 2.0"-based social networks' practices.

The collection of Big Data and the requirement for their management and analysis can potentially provide valuable insight that only such volumes can. Most importantly through, the aforementioned abundance of Big Data information as well as the existence of synergetic networks have the potential to *"shift knowledge and power from the individual to the collective"* [1] by catering for Collective Intelligence. This is achieved by addressing at least two of the three elements of the Collective Intelligence property as provided in the definition by Glenn [12]. In other words, Big Data address the element of information while the synergetic networks, that also supported Big Data, provide for Collective Intelligence's underlying interaction web for its intelligence units.

Despite the fact that Collective Intelligence is as old as humans are, its significance is by no means diminished. As Collective Intelligence refers to the capability of interconnected intelligences achieving ameliorated results in their activities than each of the single intelligences creating the collective solely would, its implications have been, are and most probably will be ever reaching. Some current well known examples of Collective Intelligence are collaboration projects such as Wikipedia[1] and the operating system Linux[2] as well as the knowledge extraction methods such as the PageRank algorithm of Google's search engine.

The advent of the domination of digitised information [14] that lead to Big Data has already had a profound effect on Collective Intelligence and is widely accepted that will continue to do so in an increasing manner [18]. The methods of analysis applied on Big Data refer to the very same behaviour that the "intelligent" part of Collective Intelligence aims to do. On the other hand, the activity of processing the Big Data by algorithms in distributed systems refers to one instance of the "collectivity" part of Collective Intelligence.

1.1 Motivation and Contribution

The identification of the overlapping segments in research for Big Data's analysis methods and Collective Intelligence's "intelligent behaviour" characteristics will provide a fertile ground for a synergy between, at least, these two domains that is necessary to be addressed. This necessity emerges not only from the *suis generis* interdisciplinary character of both Big Data's and Collective Intelligence's methods but also by the requirement to avoid reaching a state that Big Data accumulation and analysis does not lead to more intelligent insight than any individual actor.

In order to achieve the aforementioned aims, this work:

- presents concisely the state-of-the-art methods for Big Data analysis,
- provides a comprehensive account of the state-of-the-art methods for Collective Intelligence in various disciplines,
- identifies research challenges and methodological perspectives in the amalgamated domain of Big Data and Collective Intelligence.

---

[1] https://en.wikipedia.org/wiki/Collective_intelligence.
[2] http://www.linuxfoundation.org/.

The remaining of the work is organised as follows: Sect. 2 presents existing research on Big Data's key pillars, i.e. analysis methods, systems' architecture and databases solutions. Next, Sect. 3 provides a comprehensive account of the state-of-the-art methods for Collective Intelligence in prominent disciplines for varying type of actors. Finally, the work is concluded in Sect. 4 by a summary including the challenges and perspectives of the common ground between the directions of Big Data and Collective Intelligence.

2 Big Data

In an attempt to introduce Big Data based on their characteristic attributes, Laney's [16] focus rested on Volume, Velocity and Variety attributes, affectionately referred to as the "3 V's".

Volume. The increase of aggregate volume of data due to lower cost of retention as well as the perception of information as tangible asset.

Velocity. The pace at which information is created and thus requires attention (such as storage, caching, rooting, latency balancing, etc.).

Variety. The nature of the data that most often than not are in varying data formats/structures/types in addition to inconsistent semantics.

Big Data is referred to by Snijders et al. [28] as *"a loosely defined term used to describe data sets so large and complex that they become awkward to work with using standard statistical software"*. Continuing further with the attribute based definition, and as the field evolved further, more V's have been added to the initial "3 V's" in order to sufficiently address the challenges encountered, producing thus a more detailed nature of a total ten characteristic attributes (thus "10 V's"[3]) of Big Data [5].

Veracity. The requirement of access to the appropriate and enough (for training, validation, testing) data in order to be able to verify hypotheses.

Validity. The quality of the data originating from various sources based on varying schemata leading to the requirement of "cleaning processes".

Value. The business value of the data, such as Return On Investment as well as their potential to transform an organisation.

Variability. The non-static nature of data sources that lead to information that is dynamic and evolving.

Venue. The multiplicity of sources (distributed) the data originate from that make for the heterogeneity of the data.

Vocabulary. Inherent descriptors of the data such as schema, semantics and data models that depend on the content and/or the context of the data and refer to the data's structure, syntax and content.

Vagueness. Lack of clear definition of the complex and evolving term "Big Data".

---

[3] Henceforth appearing with a capital first V in order to denote the specific meaning these have for Big Data.

The collection of the aforementioned characteristics imposes a plethora of challenges to be addressed, as put by Borne [5]: *"the capture, cleaning, curation, integration, storage, processing, indexing, search, sharing, transfer, mining, analysis, and visualization of large volumes of fast-moving highly complex data"*.

2.1 Analysis Methods

Efficient analysis methods, i.e. data-driven decision making methods, in the era of Big Data is a research direction receiving great attention [30]. The perpetual interest to efficient knowledge discovery methods is mainly supported by the nature of Big Data and the fact that in each instance, Big Data cannot be handled and processed to extract knowledge by most current information systems.

The generic simplified knowledge discovery process pertains to three stages from raw data input to knowledge output. The first step is to gather from multiple sources the raw data input an perform tasks such as selection of appropriate to the process data-set, preprocessing in order to clean the data into a useful state as well as transformation to achieve common representation suitable for the next step. The second step is to perform analysis utilising methodologies of data mining, thus leading to information as an output. In the third step, information is conversed into knowledge by means of evaluation and interpretation.

Analysis methods focus on identifying hidden (due to the "10 V's") patterns, rules, associations, groupings and - in general terms - information that is new, valuable, non-obvious and very hard to get manually, from Big Data using methodologies such as artificial intelligence, machine learning and statistics.

The conversion of information into knowledge, the last step of the knowledge discovery process, is of great importance to the analysis methods of Big Data. It's the step allowing information that can only answer simplistic questions such as "who", "what", "when" and "where" to be enhanced into information that has been verified and augmented by contextual information, i.e. it's useful. Thus, knowledge allows to answer the much more complicated "how" questions [3].

Traditional data mining approaches face many challenges in Big Data [30]:

Lack of scalability. Design and implementation does not address scalability issues necessary for the Volume, Variety and Velocity of Big Data.

Centralisation. Execution is traditionally assumed to be taking place on single information systems while having all the data into the memory, a scenario that cannot be achieved for the Volume of Big Data, among other V's.

Non-dynamic attitude. The design does not tackle data that require dynamic adjustment based on analysis of the input with the Velocity of Big Data.

Input structure uniformity. The design and implementation does not cater for the large Variety of Big Data.

Bearing in mind the inefficiency of traditional data mining approaches in the face of the "10 V's", a complete redesign is required in order for these approaches to be useful for Big Data. In addition, state-of-the-art Big Data analysis focuses [8] on: (a) Inference capability based on large-scale reasoning, benchmarking and machine learning due to the Volume and Variety of Big Data, (b) Stream data processing in order to cater for the Volume and Velocity of Big Data, and (c) Use of linked data and semantic approaches in order to address challenges such as efficient indexing as well as entities' extraction and classification.

2.2   Processing Resources

The advent of Big Data with all the requirements for new analytics methods as described in Sect. 2.1, call for a new computing paradigm. A paradigm that will allow computations not to require prohibitive long time to execute while featuring disk arrays in order to hold the volume of data. Enter the Cloud and the process of computing onto the Cloud, thus Cloud Computing. The cloud metaphor refers to an obscured to the end user set of adaptable processing resources and networking that are accessible from anywhere, just like a cloud would do to engulfed items.

Current experience with Cloud Computing applied to Big Data usually revolves around the following sequence: preparation for a processing job, submission of the job, wait a usually unknown amount of time for results, receive none-to-little feedback as to the internal processing events and finally receive results, a paradigm that resembles awfully a lot the mainframe computing age [10] of thin clients and heavy back-ends. In contrast to the common processing scenario of much less "10 V"-complexity than Big Data - where one would enjoy direct manipulation, realtime interactivity and within seconds response/results - Big Data do not allow for any of these. In other words, current systems for Big Data processing offer a computing workflow that indeed reminds the 1960 s era.

Mell and Grance [21] define Cloud Computing as *"a model for enabling ubiquitous, convenient, on-demand network access to a shared pool of configurable computing resources (e.g., networks, servers, storage, applications, and services) that can be rapidly provisioned and released with minimal management effort or service provider interaction."* wherein the resources can be of physical or virtual.

Cloud Computing features a number of characteristics that make it the current choice for Big Data processing:

On-demand self-service. Resources and services are available for deployment to consumers without the need for human intervention on the provider's side.

Broad network access. Consumers can access their allotted resources and services through a variety of client platforms almost irrespectively of the clients' mobility or processing capabilities.

Resource pooling. Resources and services are pooled in a dynamic fashion in order to service all provider's customers demands, while the locality of the resource is mostly of none-to-little interest.

Rapid elasticity. The provision of resources and services is furnished and released in an elastic manner in order to scale with demand in both directions of the provider. At the same time, the consumer perceives the resources and services as limitless.

Measured service. Resources and services are provided as measurable with an appropriate model in order to manage and charge the consumer.

Moreover, Cloud Computing is provided with three key service models (Software as a Service, Platform as a Service and Infrastructure as a Service) while the deployment methods include private clouds, community, public and hybrid clouds[4].

As far as the actual software that manages the distributed processing and storage is concerned, Apache Hadoop[5] is one such popular nowadays solution. In fact, Apache Hadoop is a software framework (an ecosystem of modules) that manages clusters of (commodity) hardware, be these on the Cloud or in-premises. Moreover, in order to fully take advantage of processing and storage resources for parallelisable problems, programming models exist that profit from the locality of data, aiming at processing as close as possible to the storage thus taking advantage of reduced transit time and bandwidth use. A popular such model is MapReduce [7] that is also implemented as a module of Apache Hadoop.

Based on the aforementioned broad definition of Cloud Computing, its architecture cannot strictly be defined. The delivery of Cloud Computing resources and services usually involves numerous components that interact and feature using an intelligent interdependence methodology in order to provide for the elasticity characteristic. One such common component that is of great importance to the theme of this work is the storage/database component.

2.3   Storage

With the Volume of Big Data as well as the distributed nature of the processing resources utilised for Big data, it is no surprise that storage of such information is achieved by using mostly distributed approaches [29]. Accordingly, storage solutions for Big Data mostly refer to distributed file systems, Cloud storage, NoSQL databases as well as NewSQL databases. Traditional relational databases can indeed, in some occasions, address some of the "10 V" requirements but have been shown to be less efficient and thus more expensive [19]. This distributed character addresses both the need for Volume as well as solutions' scalability.

Cloud storage. The popularity of Cloud Computing has inevitably lead to the development of Cloud storage [32]. Cloud storage solutions usually aim at achieving as many and as high as possible of availability, reliability, performance, replication and data consistency. The service refers to both end-users as well as enterprises, while access is achieved through the internet with a variety of devices, as per the general characteristic of Cloud technologies. End-users usually store therein their personal data and backups, while enterprises' needs for large volume of information are supported with scalable, effective to capacity change and

---

[4] An extensive presentation of the service and deployment models is outside the scope of this work. Interested readers are referred to [21].

[5] http://hadoop.apache.org/.

cheap means. Cloud storage solutions usually provide for reach interfaces that cater for stored content's dissemination as well as sharing between accredited collaborators.

NoSQL databases. Currently, the predominant solution to Big Data storage, NoSQL databases primarily focus on availability, partition tolerance and speed, usually at the cost of consistency. When compared to relational databases NoSQL solutions utilise low-level query languages, lack standardised query interfaces and offer no true ACID transactions, but for few exceptions. On the other hand NoSQL solutions are of simpler design, do not require binding fixed table schemas, offer "horizontal" scaling to clustered hardware and provide fine-grained control over availability [17]. According to the data model used, NoSQL databases are categorised in key-value stores, columnar stores, document databases and graph databases.

NewSQL databases. NewSQL databases [4] are a hybrid solution of databases between NoSQL and relational databases, that features the advantages of both origins. Thus, NewSQL databases offer the transactional guarantees of relational databases in addition to the scalability of NoSQL databases. NewSQL databases have five main characteristics [31]: SQL query interface, support for ACID, non-locking concurrency control mechanism, significantly increased per-node performance in comparison to relational databases and bottleneck resistant scale-out, shared-nothing architecture when executing in many nodes. NewSQL databases are expected to be *"50 times faster than traditional OLTP RDBMS"* [29].

Distributed File Systems. Distributed File Systems (DFSs) address the management of storage in a network of systems. A wide variety of DFSs exists, though Hadoop File System (HDFS) [27] and Google File System (GFS) [11] have recently received most attention. As HDFS is part of ubiquitous Apache Hadoop, it has become the *de facto* DFS of choice offering the capability to store large amounts of unstructured data in a reliable way on (commodity) hardware while providing very high aggregate bandwidth across the cluster.

Big Data query platforms. A number of solutions exist that are in fact an interface for Big Data storage querying platforms. Despite these solutions differ as far as their underlying technologies, most provide SQL type query interfaces aiming at integrating with existing SQL-based applications. Hive, Impala, Spark, Drill, to name a few, are some of these *"SQL-on-Hadoop systems over HDFS and NoSQL data sources, using architectures that include computational or storage engines compatible with Apache Hadoop"* [2].

3   Collective Intelligence

Collective Intelligence features a plethora of definitions, the most notable of which being the one of Malone and Bernstein [18] as *"groups of individuals acting collectively in ways that seem intelligent"*. This definition presents a number of interesting characteristics, such as:

- the lack of explicit definition of the, elusive and complex, term "intelligence" catering for greater adaptability,
- the requirement of the notion of activity by the individuals, thus emphasising on the process and not the result,
- the vague definition of the individuals' grouping that allows for varying-grain individuals and group depending on the observing level,
- the requirement of collective activity by the individuals that is agnostic to their aims as long as their activities exhibit some interdependency,
- the use of the term "seems" that allows for subjective evaluation of the manifested intelligence by the observer suiting each case.

It is of interest to note that most definitions of Collective Intelligence neglect the element of the collective's orchestration. Indeed, numerous cases exist wherein the collective is formed *ad hoc* and without some for of centralised organisation, though equally in a plethora of occasions there indeed is some entity that organises the collective so as to achieve a desired or maximise an effect. In the latter case, the incentive and purpose for using resources by the organising entity is evidently focused on the amelioration of the collective's result for the profit of the organiser and the individuals. In the former case, individual entities attempt to balance the increased resources spent in order to exhibit Collective Intelligence as peer-organisation with the ameliorated, in comparison to solitary activity, output that provides advantages directly to the individuals. A third case organisation could also be assumed based on the lowest level of interdependency of the individuals' activities. In this coordination-free scenario, the seemingly unrelated individual actors' activity output are aggregated and curated [15] by a third party that selects the individuals based on their activities' collective focus.

Given the aforementioned definition's vagueness as well as broad spectrum of application, a number of domains relate Collective Intelligence [18]. Computer Science is a natural candidate as the notion of cooperating actors can easily be extended to include virtual entities (e.g. computational agents) as well. Cognitive Science and Biology are also related as the former focuses on the mind's functions that lead to intelligent behaviour while the latter on intelligent group behaviour and thus both provide for group intelligence matters. Social Sciences (such as sociology, political science, economics, social psychology, anthropology, organization theory, law, etc.) aim at behavioural characteristics of intelligent groups and thus relate with Collective Intelligence on matters of intelligent collective behaviour of complex entities with individual actors. Finally, Network Science is related with Collective Intelligence on matters where group intelligent activity is viewed on the context of networking between the group entities.

One of the key implications of Computer Science's relation to Collective Intelligence is the extension of intelligent behaviour to virtual entities. Thus, the range of individual actors is extended to human and virtual entities[6] while the collective to all combinations of these. Wikipedia is an example wherein, as far

---

[6] Collective behaviour in animals displaying intelligence attributes is established but outside the scope of this work. Interested readers are referred to Chap. 4 of [18].

as Collective Intelligence is concerned, individual actors are in their overwhelming entirety contributing humans. On the other hand, weather forecasting is an example of almost solely virtual intelligences' cooperation on producing conclusions based on sensors' input. In between these extremes, Google's search engine is an example of human and virtual intelligences' cooperation given the interaction of human intelligences' created references to websites that are intelligently processed by virtual entities to furnish the search engine's functionality [25].

Focusing on the interrelation of Collective Intelligence with Big Data, a number of research directions exhibit common research challenges and methodological perspectives. Artificial Intelligence, crowdsourcing and human computer interaction, to name a few, are of interest to the theme of this work and thus presented in the sequel.

3.1   Collective Intelligence and Artificial Intelligence

Artificial Intelligence is, in general terms, the field focusing on the intelligence exhibited by machines. The first reference appears in McCarthy et al. [20] as a *"conjecture that every aspect of learning or any other feature of intelligence can in principle be so precisely described that a machine can be made to simulate it"*. Russell and Norvig [23] organise the various definitions of Artificial Intelligence in four axes: systems that are concerned with the (a) thought process and reasoning or (b) behaviour while are evaluated based on (c) human performance or (d) an ideal concept of intelligence. The most common methodology for the deployment of Artificial Intelligence is by means of autonomous software modules (a.k.a. agents). Agents usually exhibit characteristics such as planning for future events, assuming activity based on predefined mission criteria as well as redefining said criteria based on learning from their environment.

It is thus evident that based on the aforementioned range of individual actors of Collective Intelligence, Artificial Intelligence's agents are fulfilling almost all the breadth apart from the "solely human collective" edge. Moreover, the case of amalgamated collective intelligences of both virtual and human entities is of special interest to Artificial Intelligence as well. In any case, given the numerous Collective Intelligence alternative individual actor combinations that indeed include virtual entities, the role of Artificial Intelligence to Collective Intelligence is central in one too many ways. This is also true even in cases that Artificial Intelligence's capabilities are not used as another type of intelligence in the collective activity but solely to provide for (pre)processing of data that will independently support the activities of humans.

Machine learning, the ability of algorithms to improve performance through experience with data and predict future values of data [22], has been central to Artificial Intelligence since inception and a common requirement in Collective Intelligence. Email spam filtering, recommender systems, prediction markets, machine vision, fraud detection and biotechnology, to name a few, are some of the areas that utilise Collective Intelligence and machine learning methods in order to cope with data volume and the requirement of pattern detection.

3.2   Collective Intelligence and Human Computer Interaction

The field of research for Human-Computer Interaction (HCI) deals with the design and use of interfaces that allow the bidirectional interaction between ICT and humans. Bigham et al. in [18] refer to HCI as the study of *"the links between people and technology through the interactive systems they use"* while additionally emphasise the field's interest in both the human-to-human interaction using ICT as well as their interactions online.

HCI's relate with Collective Intelligence is clearly oriented towards Collective Intelligence's individual actors range that includes humans, given the aforementioned definition of HCI. To that end, HCI's contribution to Collective Intelligence is mainly focused on allowing the collective's members to interact in meaningful and user-friendly manner with other members, both virtual and human, in order to achieve their interdependent aim. In that sense, HCI's contribution to Collective Intelligence addresses only the point of view of the individual actors. Having described the possibility of an orchestrating entity for the activity of the collective in Sect. 3, HCI can additionally support Collective Intelligence by providing ICT tools that provide incentives to the individual actors while at the same time offering methodologies for the coordination and extraction of meaning and value from the collective's activity output.

An example of Collective Intelligence assisted by special HCI characteristics is the process of the musical creation industry from idea conception to the production that features understanding and support of the musical co-creative processes dynamics. Such interfaces allow for implementation of collaboration methods and interfaces for the enhancement of the process of musical co-creation from the composition procedures up to performance and post-production levels, providing for connectivity/time/space separation of the collaborators and the artistic nature of music as well as its multifaceted creative contexts.

3.3   The Case of Crowdsourcing

Crowdsourcing, or the process of harnessing the crowd's potential in order to solve a problem, is a case of Collective Intelligence that requires special mention as it features a number of interesting, to the theme of this work, characteristics.

The original term of crowdsourcing was coined by Howe and Robinson [24] as *"outsourcing to the crowd"* but a more formal definition was provided by Estells-Arolas and Gonzlez-Ladrn-de-Guevara [9] that was the product of existing definitions' analysis, common elements extraction and establishment of basic characteristics. Their integrated solution defines crowdsourcing as *"a type of participative online activity in which an individual, an institution, a non-profit organization, or company proposes to a group of individuals of varying knowledge, heterogeneity, and number, via a flexible open call, the voluntary undertaking of a task. The undertaking of the task, of variable complexity and modularity, and in which the crowd should participate bringing their work, money, knowledge and/or experience, always entails mutual benefit. The user will receive the satisfaction of a given type of need, be it economic, social recognition, self-esteem, or the development of individual skills, while the crowdsourcer will obtain and utilize to their advantage what the user has brought to the venture, whose form will depend on the type of activity undertaken".*

Accordingly, the parties involved in crowdsourcing fall within the classes of "requestors" that is external to the group of individuals (crowd) and have an objective and the "workers" that form the group of individuals called by the requestor to address the objective.

Apart from the obvious relation of crowdsourcing with Collective Intelligence, wherein human intelligences are cooperating in order to perform a task, crowdsourcing may also be associated, even though as a special case, with the notion of the wisdom of the crowd, wherein intelligent aggregation of insight by large number of humans (crowd) is shown to be more accurate than the majority of solutions of single members' of the crowd [33].

Moreover, crowdsourcing is utilising Artificial Intelligence and HCI advances in order to achieve its function. Artificial Intelligence is used to manage both (a) the large volume of the participants (crowd), that feature a multiplicity of capabilities and skills, and (b) the task set to be allocated to participants. Accordingly, manually addressing such issues would certainly endanger efficient task-allocation management as well as quality control. As far as HCI is concerned, both classes of workers and requestors will be engaged in a human-to-human interaction that need to be addressed. Workers' contribution is greatly affected by the appropriate incentives furnished through interaction interfaces that are user-friendly and self-explanatory, while requestors necessitate appropriate interfaces in order to coordinate and harness the output of workers.

4   Challenges and Perspectives

Having described the key pillars of both Big Data and Collective Intelligence in Sects. 2 and 3 respectively, it is now necessary to address the challenges and perspectives of the shared domain of these two interrelated disciplines.

Making the most of Big Data is notoriously hard a problem, as it is common for practice for analytics to be applied with methods or on segments of data that are mostly expected to deliver results, some times in vain [26]. Bearing in mind the "10 V's" characteristics of Big Data, as described in Sect. 2, one must consider that not all "V's" need apply simultaneously in order for the data to be considered as Big Data. Accordingly, Volume and Velocity are enough criteria to label digital, clean, explicit and unambiguous data (collectively referred to as structured data) as Big Data. On the other hand, adding any of Variety, Variability, Vocabulary, Venue lead to analog, noisy, implicit or ambiguous data (collectively referred to as unstructured data), also labelled Big Data.

The human mind evolved to be able to process and make sense out of unstructured data by bringing to bear, seemingly unconsciously, an enormous amount of contextual knowledge [6]. On the other hand, computers excel at processing structured data [26]. Still, it is also a matter of processing power, not only type of data: increased Volume and Velocity of the data require more computing power of any sort. Accordingly, selection of the computing power to use on Big Data must be made based on the type of data to be processed: Collective Intelligence for unstructured Big Data, while Big Data efficient analytics for structured Big Data. A common such example is the statistical prediction task[7]: predictions will be made based on past information; virtual intelligence predictions in-line with past data are accurate; their accuracy drops for disruptive future information/events; human intelligences are surprising accurate, especially when Collective Intelligences.

References

1. Collective intelligence (2016). https://en.wikipedia.org/wiki/Collective_intelligence. Accessed 2 July 2016
2. Abadi, D., Babu, S., Özcan, F., Pandis, I.: SQL-on-hadoop systems: tutorial. Proc. VLDB Endowment **8**(12), 2050–2051 (2015)
3. Ackoff, R.L.: From data to wisdom. J. Appl. Syst. Anal. **16**(1), 3–9 (1989)
4. Aslett, M.: NoSQL, NewSQL and beyond (2011). https://451research.com/report-long?icid=1651
5. Borne, K.: Top 10 big data challenges a serious look at 10 big data vs (2014). https://www.mapr.com/blog/top-10-big-data-challenges-%E2%80%93-serious-look-10-big-data-v%E2%80%99s . Accessed 2 July 2016
6. Byrd, D.: Organization and searching of musical information (2008). http://homes.soic.indiana.edu/donbyrd/Teach/I545Site-Spring08/SyllabusI545.html
7. Dean, J., Ghemawat, S.: Mapreduce: simplified data processing on large clusters. Commun. ACM **51**(1), 107–113 (2008)
8. Domingue, J., Lasierra, N., Fensel, A., Kasteren, T., Strohbach, M., Thalhammer, A.: Big data analysis. In: Cavanillas, J.M., Curry, E., Wahlster, W. (eds.) New Horizons for a Data-Driven Economy, pp. 63–86. Springer, Cham (2016). doi:10.1007/978-3-319-21569-3_5
9. Estells-Arolas, E., Gonzlez-Ladrn-de Guevara, F.: Towards an integrated crowdsourcing definition. J. Inform. Sci. **38**(2), 189–200 (2012)
10. Fisher, D., DeLine, R., Czerwinski, M., Drucker, S.: Interactions with big data analytics. Interactions **19**(3), 50–59 (2012)
11. Ghemawat, S., Gobioff, H., Leung, S.T.: The google file system. In: ACM SIGOPS Operating Systems Review, vol. 37, pp. 29–43 (2003)
12. Glenn, J.C.: Collective intelligence: one of the next big things. Futura **4**, 45–57 (2009)
13. Hashem, I.A.T., Yaqoob, I., Anuar, N.B., Mokhtar, S., Gani, A., Khan, S.U.: The rise of big data on cloud computing: review and open research issues. Inf. Syst. **47**, 98–115 (2015)
14. Hilbert, M., López, P.: The world's technological capacity to store, communicate, and compute information. Science **332**(6025), 60–65 (2011)
15. Karydi, D., Karydis, I.: Legal issues of aggregating and curating information flows: the case of RSS protocol. In: International Conference on Information Law (2014)
16. Laney, D.: 3D data management: Controlling data volume, velocity, and variety. Technical report, META Group (2001)

---

[7] Interested readers are referred to Chap. 5 of [18] for an extensive set of Collective Intelligence forecasting examples.

17. Leavitt, N.: Will NoSQL databases live up to their promise? Computer **43**(2), 12–14 (2010)
18. Malone, T., Bernstein, M.: Handbook of Collective Intelligence. MIT Press, Cambridge (2015)
19. Marz, N., Warren, J.: Big Data: Principles and Best Practices of Scalable Realtime Data Systems. Manning Publications Co., Greenwich (2015)
20. McCarthy, J., Minsky, M., Rochester, N., Shannon, C.: A proposal for the dartmouth summer research project on artificial intelligence (1955). http://www-formal.stanford.edu/jmc/history/dartmouth/dartmouth.html. Accessed 2 July 2016
21. Mell, P.M., Grance, T.: SP 800–145. The NIST definition of cloud computing. Technical report, Gaithersburg, MD, United States (2011)
22. Provost, F., Kohavi, R.: Guest editors' introduction: on applied research in machine learning. Mach. Learn. **30**(2–3), 127–132 (1998)
23. Russell, S., Norvig, P.: Artificial Intelligence: A Modern Approach. Pearson, London (2009)
24. Safire, W.: On language (2009). http://www.nytimes.com/2009/02/08/magazine/08wwln-safire-t.html. Accessed 2 July 2016
25. Segaran, T.: Programming Collective Intelligence: Building Smart Web 2.0 Applications. O'Reilly Media, Sebastopol (2007)
26. Servan-Schreiber, E.: Why you need collective intelligence in the age of big data (2015). https://blog.hypermind.com/2015/01/28/the-role-of-collective-intelligence-in-the-age-of-big-data/. Accessed 2 July 2016
27. Shvachko, K., Kuang, H., Radia, S., Chansler, R.: The hadoop distributed file system. In: IEEE Symposium on Mass Storage Systems and Technologies
28. Snijders, C., Matzat, U., Reips, U.D.: "Big Data": Big gaps of knowledge in the field of internet science. Int. J. Internet Sci. **7**(1), 1–5 (2012)
29. Strohbach, M., Daubert, J., Ravkin, H., Lischka, M.: Big data storage. In: Cavanillas, J.M., Curry, E., Wahlster, W. (eds.) New Horizons for a Data-Driven Economy, pp. 119–141. Springer, Cham (2016). doi:10.1007/978-3-319-21569-3_7
30. Tsai, C.W., Lai, C.F., Chao, H.C., Vasilakos, A.V.: Big data analytics: a survey. J. Big Data **2**(1), 1–32 (2015)
31. Venkatesh, P.: NewSQL the new way to handle big data (2012). http://opensourceforu.com/2012/01/newsql-handle-big-data/. Accessed 2 July 2016
32. Wu, J., Ping, L., Ge, X., Wang, Y., Fu, J.: Cloud storage as the infrastructure of cloud computing. In: Intelligent Computing and Cognitive Informatics
33. Yi, S.K.M., Steyvers, M., Lee, M.D., Dry, M.J.: The wisdom of the crowd in combinatorial problems. Cogn. Sci. **36**(3), 452–470 (2012)

Author Index

Printed in the United States
By Bookmasters